Ecological Networks and Territorial Systems of Ecological Stability

László Miklós · Andrea Diviaková
Zita Izakovičová

Ecological Networks and Territorial Systems of Ecological Stability

Springer

László Miklós
Institute of Landscape Ecology SAS
Bratislava, Slovakia

Zita Izakovičová
Institute of Landscape Ecology SAS
Bratislava, Slovakia

Andrea Diviaková
UNESCO-Chair for Sustainable Development
Technical University in Zvolen
Zvolen, Slovakia

ISBN 978-3-030-06773-1 ISBN 978-3-319-94018-2 (eBook)
https://doi.org/10.1007/978-3-319-94018-2

Printed on acid-free paper

This Springer imprint is published by the registered company Springer International Publishing AG part of Springer Nature
The registered company address is: Gewerbestrasse 11, 6330 Cham, Switzerland

Acknowledgements

The publication is the result of the research supported by the grant agency KEGA Project No. 013TU Z-4/2016 and by grant agency VEGA Project No. 1/0096/1614-0735 and VEGA Project No. 2/0066/15.

Contents

Abbreviations

ABK	Abiocomplex
BSEU	Map of Bonited Soil-Ecological Unit
CLS	Current Landscape Structure
EECONET	European Ecological Network
ESLE	Ecologically Significant Landscape Element
FAR	Folk Architecture Reservation
FVZ	Forest Vegetation Zone
GIS	Geographical Information System
GTSES	General Plan of Super-Regional TSES
HPZ	Hygienic Protective Zones
ILE SAS	Institute of Landscape Ecology of the Slovak Academy of Sciences
k.ú.	Cadastral area
K_{es}	Coefficient of ecological quality of the territory
k_{odi}	Runoff coefficient index (coefficient of runoff retention for erosion models)
LANDEP	Landscape-Ecological Planning, the Methodology of Landscape-Ecological Planning
LTSES	Local Territorial System of Ecological Stability
MAB UNESCO	Biospheric reservation of the UNESCO's Man and Biosphere Programme
MCR	Memorial City Reservation
MoE SR	Ministry of Environment of the Slovak Republic
Natura 2000	The European network of protected sites of EU member states
NEAP	National Environmental Action Plan
NECONET	National Ecological Network
NFWV	Non-forest Woody Vegetation
NM	Nature Monument
NNM	National Natural Monument
NNR	National Nature Reserve

NP	National Park
NR	Nature Reservation
PA	Protected Area
PBA	Protected Bird Area
PEBLS	Pan-European Biological and Landscape Diverity Strategy
PEEN	Pan-European Ecological Network
PLA	Protected Landscape Area
PP	Natural Monument
PR	Nature Reserve
PT	Protected Tree
PZ NP	NP Protection Zone
PZ PA	PA Protection Zone
REPGES	Potential Representative Geoecosystems
RTSES	Regional Territorial System of Ecological Stability
SAC	Special Areas of Conservation, special protected area
SAS	Slovak Academy of Sciences
SEF	Socio-Economic Factor
SKŽP	Slovak Commission for the Environment
SNCSR	State Nature Conservancy of the Slovak Republic
SPA	Special Protection Areas, area of special protection
TSES	Territorial System of Ecological Stability
TSSF	Territorial System of Stress Factor
WWTP	Wastewater Treatment Plant

List of Figures

List of Tables

Chapter 1
Introduction

Territorial system of ecological stability (TSES) represents the most significant integration of landscape ecology principles into real environment policies and spatial planning practice. It has become a part of the legislation, general ecological regulation of various plans and projects, and it has become a common part of decision-making processes, as well as town-planning procedures at all hierarchical levels.

Establishing ecological networks is currently one of the main objectives of landscape ecology and nature and landscape protection (Jongman 2004). It is based on important international documents such as Agenda 21, the Convention on Biological Diversity, the Pan-European Biological and Landscape Diversity Strategy, EECONET under the European programme of IUCN, the European Landscape Convention. After all, the principles of NATURA 2000 also articulate the need to preserve the network of significant ecosystems. International policies and programmes constitute a foundation for developing national programmes of ecological networks.

However, ecological networks have several different interpretations. Since it is an attractive and seemingly easy-to-understand subject, even unqualified people without scientific education are becoming involved in it. The scientific basis for creating ecological networks is exceptionally complex and requires clarification of the functioning of the ecological stability of landscape and its significance for nature as well as man (Opdam et al. 2006). Therefore, it is vital also to explain the rudimentary theoretical and methodological principles of ecological networks.

The territorial system of ecological stability (TSES) was developed from the original type of an ecological network based on the principles elaborated by the landscape-ecological school in Brno and Bratislava, back then as a part of the complex rationalised brigade (Buček et al. 1984, 1996; Jurko 1986; Miklós 1986; Miklós et al. 1986) and other institutes. The concept of TSES was incorporated in the policy of environmental protection following the legislative changes instigated by the Slovak Commission for the Environment immediately after its founding (Miklós 1991).

L. Miklós et al., *Ecological Networks and Territorial Systems of Ecological Stability*, https://doi.org/10.1007/978-3-319-94018-2_1

From the theoretical and methodological perspective, TSES is a modern concept of nature conservation and biodiversity standing on the principles of the protection of the conditions and life forms, i.e. protection of geobiodiversity.

From the nature conservation and biodiversity point of view, the fundamental objectives of TSES include

- Maintaining a network of ecologically important segments of landscape, not only for their intrinsic ecological value, but because of their beneficial **ecostabilising effect in the surrounding landscape**, even ecologically disrupted,
- Protection of such landscape elements which, according to the traditional criteria of nature conservation, are not included in a protective category, but their maintenance **is essential** to ensure the ecological stability of the territory.

The concept of TSES can be considered a notable success aimed at incorporating the landscape-ecological principles into the legislation of crucial planning processes. Currently, the TSES is integrated into binding regulations on landscape planning, it is a mandatory part of land consolidation projects, and it has to be taken into account in water plans, flood protection documentation and the materials of environmental impact assessments. TSES is a principal knowledge concept of the current laws on nature and landscape protection. Thus, it has become a very powerful tool for ecologisation of spatial arrangement as well as functional utilisation of territory (Drdoš et al. 1995, Miklós and Špinerová 2011, Miklós et al. 2011).

Full implementation of the concept requires adopting the following measures concerning methodology, planning and projection, but also the legislation and administration of environmental protection.

As for a foreword:

The theoretical–methodical base of the TSES, as well as the procedure of TSES, has been elaborated and presented by authors of this publication in different forms in numerous theoretical publications, in methodical works as well as in projects, e.g. (Miklós 1986, 1996; Miklós et al. 1988; Miklós and Izakovičová 1997; Izakovičová et al. 2000, 2001, 2011; Diviaková 2010a, b; Diviaková and Kočická 2008; Miklós et al. 2011, *most recently* Miklós et al. 2018). *However, we consider the presented publication a complex summarising work that aggregates the theoretical knowledge, methodical competences and practical experiences gained during last decades during elaboration of numerous methodics and projects. Therefore, we deem as inevitable to present also in this publication most of the theoretical and methodical principles which built the basement of the TSES, as well as the explanation of the methodical steps of TSES, even though we repeat them good many times. Anyway we accentuate that*

- *The presented publication is the more comprehensive one comprising all the theoretical–methodical and practical side the methodics of TSES competed with extensive annex of map of real TSES projects.*
- *All textual, graphical and map works presented in the book were produced by the authors of this publication.*

References

Buček A, Lacina J, Löw J (1984) Teoretické východiská a typológia ÚSES. Pracovné materiály KRB, Agroprojekt Brno, p 12

Buček A, Lacina J, Michal I (1996) An ecological network in the Czech Republik. Veronica, Special 11th issue

Convention On Biological Diversity (1992) United Nations Conference on Environment and Development, Earth Summit, Rio De Janeiro

Diviaková A (2010a) Hodnotenie líniových formácií nelesnej drevinovej vegetácie pre potreby územných systémov ekologickej stability. VKÚ, a.s, Harmanec, p 120

Diviaková A (2010b) Nelesná drevinová vegetácia a hodnotenie jej prírodoochrannej, biotickej a krajinnoekologickej významnosti. Geographia Cassoviensis IV(1):48–53

Diviaková A, Kočická E (2008) Účinok líniových vegetačných formácií na priebeh plošnej vodnej erózie v okolí obce Žibritov. In: Izakovičová Z. (ed) Smolenická výzva IV - Kultúrna krajina ako objekt výskumu v oblasti TUR. Zborník príspevkov z konferencie, Ústav krajinnej ekológie SAV, Bratislava, pp 118–124

Drdoš J, Miklós L, Kozová M, Urbánek J (1995) Základy krajinného plánovania. Učebné texty. Technická univerzita vo Zvolene

Izakovičová Z, Miklós L, Moyzeová M et al (2011) Model reprezentatívnych geoekosystémov na regionálnej úrovni. Ústav krajinnej ekológie SAV, Bratislava, p 88

Izakovičová Z et al (2001) Regionálny územný systém ekologickej stability okresu Trnava, Ústav krajinnej ekológie SAV, 127p

Izakovičová Z, Hrnčiarová T, Králik J, Liška M, Miklós L, Moyzeová M, Pauditšová E, Ružičková H, Šíbl J, Tremboš P (2000) Metodické pokyny na vypracovanie projektov regionálnych územných systémov ekologickej stability a miestnych územných systémov ekologickej stability. MŽP SR, Združenie KRAJINA 21, Bratislava, 155p

Jongman RHG (ed) (2004) The new dimensions of the European landscapes. Springer Science & Business Media, Berlin, 258p

Jurko A (1986) Krajinnoekologická významnosť rastlinných spoločenstiev. In: Ekologická optimalizácia využívania Východoslovenskej nížiny III., ÚEBE SAV-Slovosivo Michalovce, pp 214–218

Miklós L (1986) Stabilita krajiny v ekologickom genereli SSR. Bratislava, Životné prostredie 20(2):87–93

Miklós L (1991) Zásady slovenskej ekologickej politiky. Životné prostredie 25(4):174–178

Miklós L (1996) The concept of the territorial system of ecological stability in Slovakia. In: Jongmann, RHG (ed) Ecological and landscape consequences of land-use change in Europe. ECNC publication series on Man and Nature 2, Tilburg

Miklós L, Diviaková A, Izakovičová Z (2011) Ekologické siete a územný systém ekologickej stability. Vydavateľstvo TÚ vo Zvolene

Miklós L, Izakovičová Z (1997) Krajina ako geosystém. VEDA, SAV, Bratislava, p 152

Miklós L, Lisický M, Kozová M (1988) Hrušovská zdrž a budúce ekologické podmienky okolitej krajiny. Bratislava, Životné prostredie 22(2):83–89

Miklós L et al (1986) Ekologický plán VSN. Súbor grafických výstupov. Ekologická optimalizácia využívania VSN. IV, ÚEBE SAV Bratislava - Slovosivo, p 122

Miklós L, Špinerová A (2011) Krajinno-ekologické plánovanie LANDEP. VKÚ, Harmanec

Miklós L, Špinerová A, Belčáková I, Diviaková A (2018) Instruments and tools for sustainable land use and spatial development processes. In: Belčáková I, Miklós L (eds) Landscape considerations in spatial planning processes. Peter Lang: Frankfurt a.M. et al., ISBN 978-3-631-74971-5
Opdam P, Steingröver E, van Rooij S (2006) Ecological networks: a spatial concept for multi-actor planning of sustainable landscapes. Landscape Urban Plan Elsevier 75(3–4):322 333

Chapter 2
Principles for Creating Ecological Networks

Abstract Establishing ecological networks is currently one of the main objectives of landscape ecology, nature and landscape protection. In Slovak Republic, the concept of "ecological networks" was institutionally adopted as the territorial system of ecological stability (TSES) by Decision of the Government of the SR in July 1991. TSES represents the most significant integration of landscape ecology principles into real environment policies and spatial planning practice. It has become a part of the legislation, common part of decision making, general ecological regulation of various plannings. The TSES concept is based on a geosystem approach to the landscape. This necessitates viewing the landscape space as an integrated complex of given area. The chapter characterises the functions of the primary, secondary and tertiary landscape structures for the TSES. Basic spatial units for creation of TSES considered the potential representative geoecosystems (REPGES) which are bearers of geoecodiversity, i.e. the diversity of both the conditions and forms of the life. The list of types of representative geoecosystems should serve as an ecologically based systematic framework for proposals of biocentres, as well as for other types of areas to be protected. The chapter deals also with most frequent theoretical landscape-ecological questions related to creation of ecological networks, as the stability of ecosystems and spatial ecological stability of the landscape, spatial configuration and composition of landscape. Specific attention is given to the relation of TSES to traditional nature conservation and to other types of the networks of protected areas. Concept of TSES changes the "classic" idea of the nature conservation based on the division of the landscape to protected and non-protected areas towards a system which maintains the ecological stability of the whole territory by an ecologically suitable spatial structure of the landscape even in the case that it is exploited in a different way. The TSES in Slovakia is legally defined in the act on nature conservation, and it is incorporated to the acts on territorial planning, agricultural land arrangement, watershed management, flood protection, environment impact assessment, integrated prevention and pollution control. The final part of the chapter deals with the development of the TSES and its place in the mentioned acts.

Keywords Landscape structures · Ecological stability · Geoecosystems Legislation

L. Miklós et al., *Ecological Networks and Territorial Systems of Ecological Stability*, https://doi.org/10.1007/978-3-319-94018-2_2

2.1 The Structure of the Landscape and Its Functions for Ecological Networks

The territorial system of ecological stability TSES is based on the principles of landscape ecology and represents its most successful application and implementation to the practice (Kozová et al. 2007; Barančoková et al. 2010), TSES is a spatial system, and therefore the underlying principles concerning territory, space and location must be clearly specified. Territory and space are terms which describe spatial projection and provide a spatial framework for the tangible entities of the material reality of ach spot on the Earth—the **landscape as a geosystem**.

The core aspect of the landscape as a geosystem according to the general system theory (von Bertalanffy 1968)—irrespective of numerous attempts to produce more or less accurate, more or less embellished verbal presentations—is defined by key authors as a

set of components elements of a geographical sphere and their mutual relations with each other

(e.g. Neef 1967; Chorley and Kennedy 1971; Sochava 1977; Isachenko 1981; Krcho 1991; Grodzinski 2005 and others). It should be noted that such a system, from the spatial, material and time point of view, overlaps with the content of other frequently used terms such as landscape, geoecosystems, geographical complex, but also other relational concepts such as environment, territory. Of course, the material physical section of geographical sphere is not "aware" of the name it has been assigned. According to this approach, landscape is defined as a geosystem, also in the methodology of landscape-ecological planning LANDEP (Ružička and Miklós 1982, 1990), which also includes TSES, and this definition is also adopted in the relevant laws.

The geosystem definition of landscape is as follows:

"Landscape is a complex system of space, location, georelief and other mutually, functionally interconnected material natural elements and elements modified and created by a man, in particular the geological base and soil creating substratum, soil, water bodies, air, flora and fauna, artificial structures and the elements of land use, as well as their connections, which determine also the socio-economic factors related to landscape. Landscape is the environment of man and other living organisms." (Miklós and Izakovičová 1997). This definition is confirmed by legislation, and this reading is word for word accepted in Act 50/1976 Coll. on Territorial planning and Building Code, as amended by the Act 237/2000 Coll., in §139 terms and definitions of spatial planning. The graphical presentation of this definition is shown in Fig. 2.1.

The authors dealing with the geosystem theory emphasise that all relations in the geosystems are equally important, and accordingly, the geosystem approach should not prioritise any elements. The geosystem biased studies should take into account relations of all elements with each other.

However, if considering the geosystem as the environment for living organisms (including humans) it is obvious that the geosystem is understood as a system of

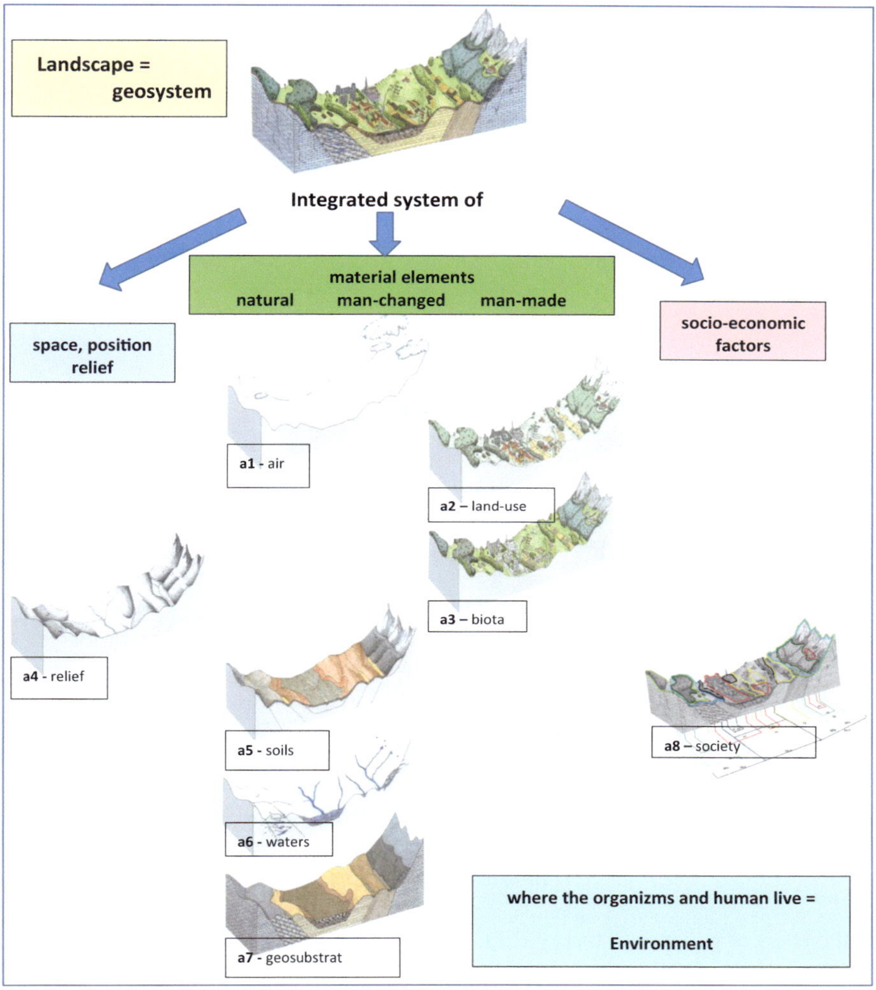

Fig. 2.1 Definition of landscape as a geosystem

conditions and forms

of the life. In this case, we centralise the living organisms—the forms of life—against all other elements of the system—the conditions of the life. The relations of living organisms to others are prioritised, they are studied in more detail. Actually, the whole environmental science in broadest sense of the word is based on this approach: if one cogitates about the "environment", there must exist something/somebody whose environment is evaluated—immediately or in further perspective.

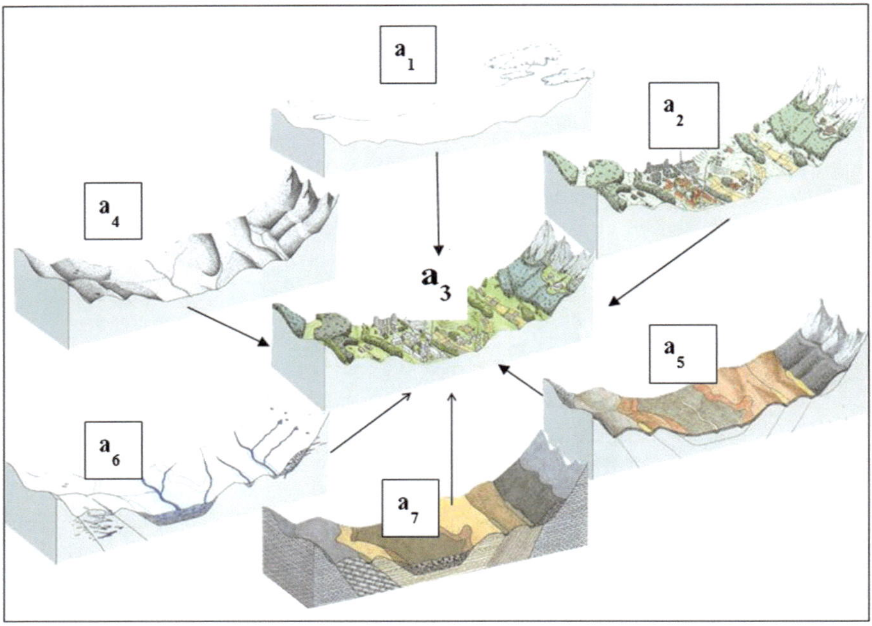

Fig. 2.2 Geosystem biased model of the ecosystem (Miklós et al. 2015)

This approach can be marked as ecological/environmental approach to the definition of the same material physical entity presented in model in Fig. 2.1a. However, centralisation of one element of this model represents actually the

model of ecosystem

on the basis of geosystem theory (Fig. 2.2).

Of course, there are many other "classic" definitions of an ecosystem in biological disciplines, e.g. based on the definition of functional compartments (Ellenberg 1973a, b; Odum 1975), but the material essence of the systems is still the same; Fig. 2.1 has the same elements as Fig. 2.2. It is to emphasise again that in reality each spot of the Earth surface is at the same time bearer of both, the geosystem as well as an ecosystem (Preobrazhensky and Minc 1973). In practical landscape-ecological studies, there is no sharp boundary between definitions of a geosystem and ecosystem approach.

The methodics of the territorial systems of ecological stability TSES concentrate on both, the conditions and the forms of the life, whereby accepts one of the basic features of the nature which reminds that if we succeed to protect the conditions we have chance to keep also the forms of life. This feature is not valid in opposite direction. This basic natural rule, i.e. the geosystem biased ecosystem approach, is the base of the development of territorial systems of ecological stability TSES.

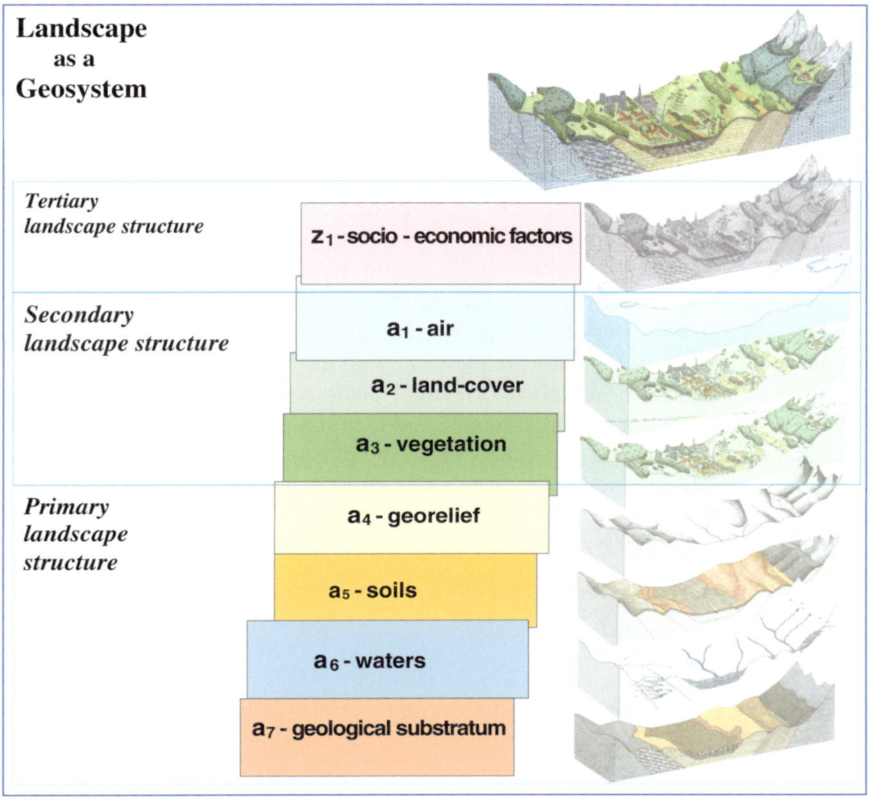

Fig. 2.3 Primary, secondary and tertiary landscape structures. *Note*: The element **a1 - air** belongs to primary landscape structure. Since physically it lies above the seondary landscape structure, it is drawn graphically on the figure in the section of secondary landscape structure

The key aspect of the geosystem approach to the landscape in its application in practical activities—herewith also in the process of designing of TSES—is the disquisitional definition of the character of its elements in terms of their functions in the construction of the whole system. Accordingly, we distinguish three partial structures—the **primary, secondary and tertiary landscape structures** with different physical material character and different functions for the spatial planning process (Drdoš et al. 1995; Miklós and Izakovičová 1997; Hreško et al. 2003; Renetzeder et al. 2010a, b; Špinerová 2015). Ecologically optimum spatial organisation and use of the landscape—which is also one of the main goals and principles of the TSES—must accept and harmonise the decisions with the features of all three partial structures (Fig. 2.3). The brief characteristic of these structures is as follows:

- **Primary landscape structure** (particularly its abiotic elements)—these conditions are physically more volatile, material and structural basis is very hard or

impossible to change (e.g. limestone is not granite, where mountain ranges are, there are not lowlands, cold climate is not warm climate), the principles of their behaviour are unchangeable or partially changeable, and their reaction to disturbance is difficult to control. These elements have in practice been least changed by mankind in comparison with the secondary and tertiary structures of the landscape, which mankind has directly created.
- **Secondary landscape structure**—human-made or altered landscape elements (mainly current vegetation and animal communities, anthropogenic structures and materials, elements of land use)—these conditions are physically bound to certain locations, and changing them requires exerting a considerable amount of energy. Nowadays, the secondary landscape structure is also defined as the current landscape structure as its elements are identical with the land-use elements.
- **Tertiary landscape structure** (mainly socio-economic factors and processes)—these do not physically exist, and they manifest in various regulations, laws, standards and legal constraints including plans, strategies, agreements, conventions, legally declared zones, categories of protected areas.

2.2 Stability of Ecosystems and Spatial Ecological Stability of the Landscape

As already mentioned, the theoretical–methodical base of the TSES, as well as the procedure of TSES, has been elaborated and presented by authors of this publication in different forms in numerous theoretical publications. However, we deem as inevitable to present also in this publication most of the theoretical principles which built the basement of the TSES.

The main theoretical cornerstones for the development of the TSES methodics were the followings:

Landscape ecology is nowadays a broadly developed scientific branch with many traditional and new mainstream concepts (Forman and Godron 1986; Richling and Solon 1993; Cook and van Lier 1994; de Groot et al. 2010, Antrop 2013, Kiss and Szabó 2016). However, the concept of ecological stability because of its intangibility remained a core question probably also for next generations of scientists. Of course, there is a mass of studies approaching the concept but outcomes from whatever research evokes always newer questions.

What is important from the perspective of applied researches, herewith also for the concept of the territorial system of ecological stability TSES? Since we started with the idea that TSES should help to keep both the conditions and forms of the life, including the maintaining the diversity of conditions and forms, let us start with the generally accepted ecological principle that the species or communities are considered threatened and their stable functioning can be impaired or rendered impossible, i.e. the organisms, populations, communities, as well as their development and diversity are endangered if

- Their **living conditions** are unfavourable.
- They are **spatially isolated** (e.g. Odum 1975, Szabó 2008).

It follows that if we wish to maintain the stable existence and functioning of diverse living systems in the landscape, we have to ensure both suitable living conditions and **spatial interconnectivity** of these systems.

What is the implication of this principle in concern with TSES? In the real world with current stage of the society, the humans need a variety of ecosystems—environmentally stable, insufficiently stable and unstable man-made, modified agroecosystems—as well as newly created systems with settlements, transport and industry. These are present in specific areas of the landscape, in the form of the elements of the current landscape structure as a result of the exploitation of territory (hereinafter "the elements of CLS") (Ružičková and Ružička 1973; Miklós and Izakovičová 1997; Szabó 2007, 2008). In terms of the aforementioned theories, man seeks to maintain the ecological stability of the landscape, **as a whole**, on a certain conventional modelling level even with the existing mosaic of various ecologically stable as well as unstable elements of CLS (Ahern 2002). The prerequisite for maintaining such "stability" in addition to the **internal ecological quality** (internal stability) of the most stable elements of the CLS is their **interconnectivity** across the territories made up of different elements with different, even lower internal stability (Grodzinsky 1987). Naveh and Lieberman (1993) argue that we should achieve a landscape which might be locally unstable, but globally stable, and it is clear that such a landscape model cannot be attained simply by protecting only isolated—even stable—ecosystems, but primarily by instituting a spatial system of interconnected, ecologically stable elements of the CLS in the frame of real, man influenced and man-created elements of the landscape. This principle is elaborated in a mass of scientific studies (Forman and Godron 1981; Mander et al. 1987, Agger and Brandt 1988; Csorba 1989; Forman 1990; Míchal 1992, Grodzinski 2005). Recently, the ability of the ecological stable elements to provide ecological stability for the whole landscape is defined in the mainstream landscape-ecological concept as the ecosystem service (de Groot et al. 2010, Bastian et al. 2015)

The stability of ecosystems depends on its intrinsic ecological quality, and consequently, the spatial stability of the landscape depends on the amount and structure of ecosystems with different intrinsic stabilities (Ružička et al. 1982a, b; Csorba 1989; Míchal 1992, Miklós et al. 2018). Albeit the differences in wording of the definitions of ecological stability, a commonly acceptable characteristic of the ecological stability is "the ability of ecological systems to withstand a period of disturbance and reproduce their essential characteristics in conditions of external disturbance" (Míchal 1992). This includes many other aspects as **resistance** (minimal change during disturbance) or **resilience** (spontaneous recovery to its initial condition). Also, Buček and Lacina (1995)—the pioneers of the TSES method—when developing the methodics of TSES defined the issue in a purpose-oriented way with outlook to use these principles in TSES projecting, as:

- The internal (endogenic) ecological stability of the ecological system is its ability to subsist under normal influence of environmental factors, including those to which the ecosystems have adapted. It is determined by the strength and quantity of the internal connections in the ecosystem. High internal ecological stability is found mainly in successional mature ecosystems (naturally occurring).
- External (exogenic) ecological stability is the ability of the ecosystem to resist effects of extraordinary factors, to which the ecosystem has not adapted through its natural evolution (extreme temperatures, widespread fires, earthquakes, volcanic eruptions, air pollution, water regime changes, chemicalisation of agriculture, etc.) (Buček and Lacina (1995).

According to all theoretical knowledge and purpose orientation to TSES, we can define the **spatial ecological stability** of the landscape as an extent of dynamic landscape ability to maintain its vertical and horizontal relations at an acceptable conventional (modelling) level, providing the landscape is composed of ecosystems with various, even low level of ecological stability (Miklós et al. 2011, 2018). Based on all above said, we can state the main purpose of TSES as a specific concept of ecological networks as:

The main purpose of the establishment of ecological networks is preserving the spatial ecological stability of the landscape

2.3 Spatial Configuration and Composition of the Landscape

The original core issue of landscape ecology is its spatial ecological structure (Tansley 1935; Forman and Godron 1986; Richling and Solon 1993; Grodzinski 2005). The authors have conducted research into spatial relations of ecosystems, spatial arrangement and spatial impact of ecosystems on their surroundings. Such scientific investigation strives to demonstrate the significance of ecologically stable landscape elements due to their spatial influence. Spatially interconnected, ecologically stable landscape elements are generally referred to as **ecological networks**. Methodologies and approaches to their study are flourishing throughout the world (Jongman 1995, 2008; Wrbka et al. 2005; Fabos 1996; Sepp and Kaasik 2002; Jongman et al. 2004; Brandt 1995; Cook and van Lier 1994; Miklós et al. 2018) and also include the methodology of TSES.

There are developed several theories—e.g. chorologic approach to the landscape, island biogeography, fragmentation theory, definition of metapopulation (Neef 1963; Macarthur and Wilson 1967; Buček and Lacina 1984; Brandt 1985; Opdam et al.

1995; Hanski 1998; Csorba 2008; Špulerová et al. 2013; Hellmund and Smith 2013)—which created a cornerstone for development of ecological networks, herewith also for TSES.

One of the best developed is the theory of island biogeography. Fragments of different greenery—shrubs, woods, wetlands—in agricultural land serve as "islands" for maintaining the stable ecosystems. But what is inevitable for ecological stability is to ensure the interconnections between those islands. The existence of correlation between the size of a natural element, shape and distance between islands is proved by many studies (Forman and Godron 1981; Agger and Brandt 1988; Solon 1988; Moyzeová and Grotkowská 2006). The surrounding landscape also plays an important role as it is prone to change induced by anthropogenic activities. Major factors affecting species diversity of these "islands" are the size of the habitat fragment (direct correlation) and the degree of isolation (negative correlation). Ecological networks are designed to preserve the minimum area required for survival of natural communities or target vulnerable populations and mitigate the isolation effects resulting from the isolation of landscape elements in the anthropogenic landscape, for example, by means of designating biocorridors, and, in extreme cases, by technical solutions, such as ecoducts (Schreiber 1988; Míchal et al. 1991; Konkolyné Gyuró 2005; Kuttner et al. 2013; Miklós et al. 2018).

The metapopulation theory stated that fragmented habitats inhabited by smaller subpopulations are threatened by the same phenomena as human populations—-natality, mortality, migrality, etc. The communication between subpopulations is of utmost importance. The results address the necessity of preserving or restoring interconnectivity of landscape elements in the current landscape and a possibility of replacing locally extinct sub-population (Tillmann 2005; Miklós et al. 2018).

Fragmentation of habitats affects various types of organisms differently. The metapopulation theory explains the impact of fragmentation on different populations (Csorba 2008). Increasing anthropogenic pressure in fragmented agricultural landscapes results in reduced size and quality of patches and increased resistance (eliminating of hedgerows), which consequently upsets the balance between the rate of extinction and recolonisation, to the extent where, on average, fewer patches are colonised (Špulerová et al. 2011, 2013; Jongman 2002). As the proportion of empty patches grows, the probability of species subsistence decreases. Current trends and research confirm the relevance of the theoretical basis for the concept of ecological networks in the landscape (Bennett and Mulongoy 2006; Štefunková 2011, Schilleci et al. 2017, Miklós et al. 2018).

A coherent spatial structure builds on the approach of Forman and Godron (1986, 1993), who promote the evolutionary theory of patches and the landscape matrix which can be subsequently employed in planning and designing of corridors.

Corridors and patches are important and ubiquitous landscape features (Brandt 1985; Kubeš 1996; Moyzeová 2010). Their origin and character are affected by disruption, heterogeneity of landscape resources and anthropic impacts. Finally, the process results in exceptionally diverse species dynamics, stability and change (Csorba 1996; Hilty et al. 2006). The key aspect of corridors is their interconnectivity or the presence of gaps. The patch-corridor-matrix paradigm introduced by Forman and

Godron (1986, 1993) enhanced understanding and analysis of the landscape as a whole in terms of the developmental dynamics of its individual components. The theory allows the analysis of landscape components as well as the relations existing between them (Štefunková and Hanušin 2015; Miklós et al. 2018).

2.4 Ecological Networks

The concept of TSES is closely related to the theories and practices on building ecological networks. It develops and applies to the practice of such concepts as the Pan-European Biological and Landscape Diversity Strategy (PEBLS), which paved the way for the Pan-European Ecological Network (PEEN), as well as the concepts and strategies of the National Ecological Network, NATURA 2000—the Birds Directive, the Habitats Directive (EEC Directive No. 79/409/EEC and 92/43/EEC), agroenvironmental programmes and specific methodologies such as concept of representative geoecosytems, mapping of grasslands, green infrastructure (Miklós et al. 2006).

The Pan-European Biological and Landscape Diversity Strategy constituted the first comprehensive coordinating framework for all nature conservation activities in Europe. "*The Pan-European Biological and Landscape Diversity Strategy is a European response to support implementation of the Convention on Biological Diversity. The Strategy introduces a coordinating and unifying framework for strengthening and building on existing initiatives. It does not aim to introduce new legislation or programmes, but to fill gaps where initiatives are not implemented to their full potential or fail to achieve desired objectives*" (Pan-European Biological and Landscape Diversity Strategy 1996). It motivated a number of concepts and projects, e.g. the concept of maintaining "green infrastructure" in Europe. The principles of PEEN were implemented at a regional and local level in most of the European countries (Mabder et al. 1987; Rientjes and Drucker 1996; Tardy and Végh 2006).

However, the understanding of ecological "networks" is very different. Are all "networks" really networks? The phenomenon referred to as **ecological networks** can be grouped into three major categories (Miklós et al. 2011, 2018):

(a) An ensemble of nature conservation areas of "**natural heritage**" character. This is a set of traditionally protected areas, typically the most natural and best-preserved areas, rare habitats, curiosities, national natural peculiarities, endemic factors. It is a cultural, natural, historical approach to ecological networks of the character of the natural heritage. Such areas form a significant experimental basis for ecological research. They are not interconnected, and the spatial relationships of these areas are irrelevant; therefore—in the strict sense—they are not really networks. Such a "network" is built up by the oldest national conservation areas, which include the network of ***European Diploma Sites*** (1965) and the ensemble ***Biosphere Reserves*** of UNESCO Man and Biosphere Programme dating from 1976. These are the initial activities of the Council of Europe in the field of ecological networks.

All these initiatives substantially contribute to the maintenance of the ecological stability of the landscape. Nevertheless if analysing the criteria, e.g. for European Diploma programme, they prioritise cultural-natural heritage aspects, as representativeness, historic, aesthetic, scientific or recreational value determined by the particular characteristics of fauna, flora, geology, climate and geographical situation, and also protection status (Jongman 1995). Similarly, the European network of Biosphere Reserves is the basis for the cooperation of the Member States of the Council of Europe in the field of nature protection. The selection is also based on the criteria of: the value for nature conservation (unique, typical and endangered) and current protection status.

(b) Network of various protected "**landscape types**". It is a physical (bio)geographical approach to the establishment of ecological networks. The aim is to preserve as many representative landscape types as possible. However, the focus remains on the natural or endangered landscape types, the spatial interconnectivity is not a criterion, and in reality, they are not always in the form of spatially interconnected systems. Anyway, this kind of "networks" is the most commonly employed and widespread worldwide and also includes the policy NATURA 2000.

(c) Spatial ecological systems are in reality spatial networks aimed at preserving both the inner functions and spatial relationships in the whole range of ecosystems (not only the natural ones); i.e. the emphasis is put on maintaining spatial relations and not on the quality of selected ecosystems or territories. It is **a landscape-ecological approach** to the creation of ecological networks. This approach forms the basis for modern concepts of **territorial systems of ecological stability**.

> Any set of valuable nature reserves does not automatically constitute an ecological network, and it becomes one only after interconnectivity of the reserves has been established.

Ecological networks and nature conservation

It should be noted that the landscape-ecological approach does not contradict the two "classic" approaches. On the contrary, nature reserves are the pillars of the territorial system of ecological stability. The new approach ensures that the principles concerning the persistence of the conditions and life form apply across the whole territory and in all types of ecosystems. Furthermore, the classical approaches may lack the desired complexity and full-territory aspect simply because they do not promote integrated networks, but only a set of territories instead.

This generated a lot of controversy among experts regarding the new way of nature conservation. The basic aspect of the debate can be delineated by the following questions:

(a) What is "**nature**"? only a landscape segment in a model state, which is conventionally referred to as "natural state", or a natural part of the landscape, or the landscape as a whole, as a geosystem or the geosphere?

Admittedly, the landscape-ecological approach defines nature, for conversation purposes, as a whole territory, a wide array of landscape components which also make up nature and foster diversity of conditions and various life forms.

(b) What is the objective of "**nature protection**?" protection of only a selected landscape segment, or the protection of the whole landscape? The landscape-ecological approach adheres to the whole-territory landscape principle and highlights maintaining the functions within various ecosystems (not only in their natural state) and spatial relations, exchange of matter, energy and genetic information between ecosystems in the landscape.

Therefore, the new concept of nature and landscape protection in the Slovak Republic concentrates on a gradual transition from "black-and-white" division of the landscape, which recognises only protected and unprotected landscape areas, to a whole-space covering system of maintaining ecologically sustainable landscape structure and spatial ecological stability of the landscape with a differentiated way of use (or non-use) of the territory. This is the central purpose of systems of ecological stability (Jongman and Pungetti 2004; Fabos 1996).

The above-mentioned theoretical–methodological and practical principles lead to the development of TSES as a practical and legally binding tool of a modern whole-territory landscape management in compliance with the provisions of the Convention on Biological Diversity AGENDA 21 from the UN Conference on Environment and Development in Rio Summit in 1992.

2.5 Potential Representative Geoecosystems

Identification and characterisation of individual types of potential representative geoecosystems (REPGES) are of **strategic importance** for nature conservation as well as the development of TSES (Miklós et al. 2006). It should serve as a framework scheme for a systematically structured plan to determine:

- What representative conditions and forms of geoecosystems are present in a given territory.
- And thus what needs to be protected and preserved.
- Which are actually preserved and protected, and the resulting findings.
- Which are preserved without being protected.
- Which are and which are not physically preserved in a given territory.

It is also necessary to devise a **strategy for establishing and protection of biocentres** in each territory.

This approach is of higher quality than the conventional practice based on searching for the best-preserved habitats according to their current structure and bolsters the arguments in favour of preserving certain landscape segments more than the "classic" conservationist argumentation does. From the professional point of view, this concept encourages the preservation of geoecodiversity, which is a vital prerequisite for biodiversity conservation.

It should also be noted, however, that so far, for a variety of reasons, the mentioned "classic" approach is more widespread, primarily because of misinterpretation of this concept, incompetency of some members of TSES project teams, but also various subjective reasons.

The following are the methodological notes concerning this concept.

2.5.1 Geoecosystems—An Object of Geoecodiversity Conservation

Geoecosystems are particular objects and bearers of **geoecodiversity**.

The majority of geographers and geosystemologists define the term "geosystem" as a sufficiently comprehensive representation of the landscape sphere. On the contrary, numerous natural science disciplines and the general public may associate the prefix "geo-" entirely with inanimate elements of landscape sphere, or even with the geological substrate. Therefore, in order to eliminate the ambiguity a prefix "geoeco-" can be used to denote the terms which stress both aspects of the indivisible landscape sphere—the abiotic and biotic systems.

The starting point of the protection strategy of diversity of conditions and life forms is the identification of the geoecosystems that need to be preserved (Bailey 1995; Bunce et al. 1996). If we are to tackle the issue in depth, we would observe that virtually every inch of the landscape sphere provides certain living conditions for organisms living there, and humans have no right to impose inferiority on any species (e.g. wild solanaceous weeds, pioneer species on tailing piles or rodent populations in towns and cities) and ultimately deny their protection or even destroy them.

In practical terms, such in-depth ecological approach may seem inapplicable, and therefore, these efforts stem from certain geoecosystem models, which—underpinned by a general scientific consensus influenced by cultural, historical, ecological, biogeographical and regional aspect—identify the worth-preserving geosystems, geosystems which deserve preservation or geosystems which, from the anthropocentric point of view, do not deserve protection (or are even considered harmful).

Therefore, to preserve geoecodiversity, we develop the models of potential representative geoecosystems.

A potential representative geoecosystem (REPGES) is defined as a framework, hierarchically higher geoecosystem, which provides the most representative characterisation of the landscape as a geoecosystem in a given territory and **hierarchical level**.

Such identification and definition of REPGES should serve as the basic strategic framework for the selection of worth-preserving and needful-to-preserve landscape segments.

Such identification and characterization of REPGES subsequently serve as a framework for the selection of the landscape segments which are worth preserving and those requiring preservation.

2.5.2 *Identification and Characterisation of Potential Representative Geoecosystems*

Establishing criteria for determining representative ecosystems poses a complex issue (Jongman 1996; Bunce et al. 1996). "**Representativeness**" envisages accurate determination of the hierarchical significance level of a certain geoecosystems, which can be subjectively misjudged as illustrated by a case of a wide floodplain which gradually extends from a basin to a valley and ends in highlands. There is no doubt that in the basin the floodplain is a representative geoecosystem dominating in terms of function and area. However, can be a floodplain considered a representative geoecosystem also in the valley, and if so, to what extent? Representativeness should also provide a **framework**—dominating, typical geoecosystem—which allows its lower-level hierarchy segmentation and delineation into various other geoecosystems. There are a number of such instances, e.g. the occurrence of upland geoecosystems on highland plains or cliff geoecosystems in lowlands.

These issues are addressed by the determination and characterisation of potential representative geoecosystems in the whole territory of Slovakia respecting certain hierarchical level, precision and objectivity (Miklós et al. 1997, Kočická 2011, Kočická and Kočický 2014). This organisation provides a framework for lower hierarchical levels which allow more detailed classification (Bezák et al. 2010).

The term "**potential**" means that

- Identification of REPGES is, at the first level, based on the assessment of the primary landscape structure—abiotic conditions—conditions for the development of potential natural vegetation, whereas
- The second level relates to the potential natural vegetation which would develop in a particular territory and considers the biotic conditions of the territory, even though a completely different type of real vegetation prevails at the given point in time.

The basic output of REPGES is to determine:

- Which type of a REPGES is representative—characteristic for a region in question.
- Which region is characterised by a certain type of a REPGES.

At the same time, it is crucial to entertain the following hierarchical principles

- Each type of geoecosystem is representative somewhere.
- Each region has a representative geoecosystem.

The methodology of identification and characterisation of REPGES is described in the several works as Miklós et al. 1997, Izakovičová et al. 2011. The application of REPGES to the project of TSES is obligatory step of methodics issued by the Ministry of Environment of SR (Izakovičová et al. 2000), confirmed also by legislation (Decree of the MoE SR No. 24/2003 amending the act on nature and landscape protection) specialist from the landscape-ecological schools in Brno

2.6 Concept of the Territorial System of Ecological Stability and Its Development in Slovakia

The concept of ecological networks and its application in the planning practice in former Czechoslovakia and later in Slovakia was formulated as the **territorial system of ecological stability TSES**. It can be stated that this concept, as well as its implementation to the practice, was ahead of its alternatives produced in Western Europe and it offers one of the most comprehensive concepts of this kind. It represents a linear interconnection of the landscape-ecological research, applied research, legal incorporation into the laws governing spatial arrangement of the landscape and methodologies, routine planning and implementation of TSES in the landscape.

2.6.1 Development Milestones of the Territorial System of Ecological Stability

The concept of TSES in Slovakia observes the original principles elaborated in cooperation of specialists from landscape-ecological schools in Brno (Buček et al. 1984, 1996) and Bratislava from the Institute of Landscape Ecology of the Slovak Academy of Sciences as a part of the so-called rationalising brigade (Miklós et al. 1985, 1986). TSES was incorporated into first official environmental policy issued by the Slovak Commission for the Environment (Miklós 1991).

TSES is globally perceived as one of the most advanced concepts aimed at the protection of nature and biodiversity. Complete TSES projects on national level, including TSES terminology, were developed in Slovakia as a part of the so-called General Ecological Model of the Slovak Socialist Republic (in Slovak Ekologický general SSR, Miklós et al. 1985), further developed by numerous projects as for ecological assessment of the Hrušov Dam (1986), the greening project of Slovakia (1990) and other. TSES formed an integral part of comprehensive proposals within the methodics of the landscape-ecological planning LANDEP (Ružička and Miklós

1982, 1990) which underscores the whole territorial aspect. These projects, however, were not legislatively underpinned until 1989.

After 1990, the TSES concept became one of the central programmes of the Slovak Commission for the Environment in the spirit of the preparatory conferences for Rio Summit 1992. The TSES conception adopted by Government Resolution of the Slovak Republic No. 394 of 23 July 1991 and the General Plan of Supra-Regional TSES by Government Resolution No. 319/1992 paved the way for the incorporation of the TSES principles into the environmental policy and planning. The government resolutions constituted the basis for the implementation of TSES into the legislation of other government departments (see below) as well as other TSES-related activities. The whole process was backed up by various associated projects (identification of ecologically important areas, unified GIS materials and other). The concept of TSES became one of the central programmes of the Slovak Commission for the Environment in the spirit of preparatory conferences for Rio Summit 1992.

The TSES projects were launched in 1991, after the development and approval of the concept of TSES. Drafting of the TSES projects in Slovakia was performed on a "top-down" basis—from the General Plan of supra-regional TSES, through regional to local TSES, in accordance with the levels of spatial planning documentation.

In 1992, the first draft of the General Plan of super-regional TSES (GTSES) was developed for the territory of the Slovak Republic (Húsenicová et al. 1992). It is a basic framework for spatial arrangement of the most ecologically important natural territories and establishes relations and status of ecologically stable territories in Slovakia in terms of their linkage to the European system of ecologically stable territories, which ultimately produced a crucial document for the protection strategy of ecological stability, biodiversity and genetic resources of the SR (SKŽP SR 1992). GTSES was elaborated at a scale of 1:500,000 and 1:200,000 and was approved by Government Resolution No. 319 on 27 April 1992.

Following the General Plan, the National Ecological Network (NECONET) was prepared (Návrh národnej ekologickej siete Slovenska–NECONET 1996). The NECONET was based on the concept of the European Ecological Network (EECONET) and drew on the principles of the Dutch National Ecological Network (Bennett 2002; Jongman and Kristiansen 2001; Tardy and Végh 2006; Bonnin 2007; Jongman et al. 2011). It represents a network of important, especially protected areas significant for the conservation of genetic resources and biodiversity. It focuses on delineation of core areas (similar to biocentres in TSES), ecological corridors (similar to the corridors in TSES) and development of natural elements of European and national significance with the aim to establish an integrated system of protected areas and potentially valuable territories in individual European countries observing international criteria and standards. In Slovakia, NECONET was developed in 1996.

Further, milestones of the implementation of TSES of that period include:

- Legal framework: Act No. 330/1991 Coll. on land consolidation, land property arrangements, Act on Territorial planning No.262/1992 Coll. (amendment), Slovak National Council Act on Nature and Landscape Protection No. 287/1994 Coll.

This act also defines establishing and maintaining the TSES as a general **public interest**.

- Methodology: The methodology of TSES and the General Plan of super-regional TSES of the SR (Húsenicová et al. 1992), methodological guidelines (1993) and various manuals (1993–1995, e.g Jančura et al. 1994). The conception was bolstered by other major projects—habitat mapping, remote sensing materials and GIS.

The regional TSES (RTSES) was developed in the years 1993–1995. The RTSES was devised for all 38 regions (former districts of Slovakia) at a scale of 1:50,000, in some instances, 1:25,000 and 1: 100,000.

Following the preparation of the regional TSES in 1993, the processing of the local TSES (LTSES) was launched; however, it was not performed in such a coordinated manner as the RTSES. The only case of coordinated development of the LTSES was initiated by the Ministry of Environment of SR in 1996–1997 in selected cadastral territories within the reach of ZSNP Žiar nad Hronom. Other LTSES projects were prepared as a part of spatial planning documentation (Moyzeová 2010).

In 2000, the methodological guidelines for the development of the regional and local TSES were formulated (Izakovičová et al. 2000). The GNTSES was updated in 2002 (Miklós et al. 2002a, b) as was required by the National Environmental Action Plan (NEAP) approved by Government Resolution No. 350/96, in which Sector E—nature and landscape conservation and spatial development adopt measures to "unify the projects of the regional TSES with the aim of updating the GTSES of the SR".

A major milestone of nature conservation was Act No. **543/2002 Coll**. on Nature and Landscape Protection as amended, which defines the TSES, and in the §3, Section 3 stipulates that the development and maintenance of TSES are a public interest.

As Slovakia joined the EU, it also adopted the European system of nature protection based on the system of protected areas codenamed **NATURA 2000**. The main objective of the network of protected areas of the EU Member States is to preserve natural heritage significant not only for the Member States, but also for the entire EU. This system of protected areas aims to afford the protection of the rarest and most endangered species of wild flora and fauna as well as natural habitats in all biogeographical regions in the territory of the EU Member States, and by protecting the habitats of these species, to ensure the preservation of biodiversity throughout the European Union. The establishment of the NATURA 2000 network is underpinned by the following EU legal norms:

- Directive 2009/147/ES of the European Parliament and Council (codified version of the original Directive 79/409/EEC adopted in 1979) on the conservation of the wild birds (also known as the Birds Directive),
- Directive 92/43/EEC on the conservation of natural habitats and of wild fauna and flora (also known as Habitats Directive).

According to these directives, the NATURA 2000 network consists of two types of territories:

- Special Protection Areas (SPAs)—designated under the Birds Directive—in national legislation: protected bird areas,
- Special Areas of Conservation (SAC)—designated under the Habitats Directive—in national legislation: areas of European importance—before designation as an SAC; after the area is designated as a SAC, it is integrated into the relevant national category of protected areas.

These two directives represent the most comprehensive legal standard for the protection of nature in the world. The SPAs and SACs are obligatory parts also of TSES projects.

The strategic principle of the TSES approach was an innovative **whole-territory** approach to the nature conservation, conservation of the diversity of both the condition and form of the life—i.e. the **geoecodiversity**—removing spatial gaps, balancing the stability of ecosystems and stability of the landscape. Despite, or maybe even because of this, since it was adopted, the TSES has come in for a lot of scrutiny and doubt from environmentalists as well as the general public. The complete original form of the TSES concept did not win over the professionals from the Nature Conservation Department or other departments, and only the partial aspects of TSES were presented—mostly those regarding delimitation of biocentres and biocorridors, which gradually became fashionable. However, for example, the designation of ecostabilising measures in unprotected areas was (and maybe still is) not well received by many environmentalists. Only a few people who were involved in the drafting process of the concept and its methodology were willing to acquaint themselves with the original and complex concept. Thus, we consider it essential to address this shortcoming and, respecting the existing objectives, try to remedy it by means of publishing the key principles of the original concept.

2.6.2 Implementation of the Territorial System of Ecological Stability into the Legal System of the Slovak Republic

The concept of the TSES can be considered the most notable and evident success of incorporating the landscape-ecological principles into the legislation of crucial planning processes.

The real contribution of the TSES concept is its incorporation through the relevant policies and laws into the spatial planning processes at the outset of the adoption of new environmental legislation in the 1990s. They included the following laws:

(a) ***Act of the Slovak National Council No. 330/1991*** on land consolidation, land property arrangements, land offices, land pool and land associations as amended in later regulations (in particular No. 293/1992 Coll., 323/1992 Coll., 187/1993 Coll., 180/1995 Coll., 222/1996 Coll., 80/1998 Coll., 256/2001 Coll., 420/2002 Coll., 518/2003 Coll., 217/2004 Coll., 523/2004 Coll., 549/2004 Coll., 571/2007 Coll., 285/2008 Coll., 66/2009 Coll., 499/2009 Coll., 136/2010 Coll., 139/2010 Coll.,

559/2010 Coll., 547/2011 Coll., 345/2012 Coll., 145/2013 Coll., 180/2013 Coll., 115/2014 Coll., 363/2014 Coll., 122/2015 Coll., 125/2016 Coll.) provides that:

- The Territorial System of Ecological Stability is a mandatory part of projects of land consolidation.
- The elements of the TSES and important landscape elements are considered societal facilities in the context of land consolidation.
- "Ecostabilising measures" are a part of the TSES especially in case of agricultural land (Hrnčiarová, Izakovičová et al. 2000).

(b) ***Act 50/1976*** Coll. on the Territorial Planning and Building Code as amended in later regulations (in particular No. 103/1990 Coll., 262/1992 Coll., 136/1995 Coll., 199/1995 Coll., 286/1996 Coll. 229/1997 Coll., 175/1999 Coll., 237/2000 Coll., 237/2000 Coll., 416/2001 Coll., 553/2001 Coll., 217/2002 Coll., 103/2003 Coll., 245/2003 Coll., 417/2003 Coll., 608/2003 Coll., 541/2004 Coll., 290/2005 z. of., 479/2005 Coll., 24/2006 Coll., 218/2007 Coll., 540/2008 Coll., 66/2009 Coll., 513/2009 Coll., 118/2010 Coll., 145/2010 Coll., 547/2010 Coll., 408/2011 Coll., 300/2012 Coll., 300/2012 Coll. 180/2013 Coll., 219/2013 Coll., 368/2013 Coll., 293/2014 Coll., 314/2014 Coll., 154/2015 Coll., 247/2015 Coll., 254/2015 Coll.). In 1990, spatial planning came under the Ministry of Environment. As a consequence, the law "turned greener" through amendments (Act No. 262/1992 Coll., and Act 237/2000) which define:

- Ecologically optimum spatial arrangement and functional utilisation of territory (landscape-ecological plan) are an obligatory regulation for utilisation and spatial arrangement of territory.
- TSES or its elements (biocentres, biocorridors, interactive elements, ecostabilising measures) also defined as an obligatory regulation.

(c) ***Act No. 287/1994*** and subsequent ***Act 543/2002*** on Nature and Landscape Protection as amended (in particular r1/c48/2003 Coll., 525/2003 Coll., 205/2004 Coll., 364/2004 Coll., 587/2004 Coll., 15/2005 Coll., 479/2005 Coll., 24/2006 Coll., 359/2007 Coll., 454/2007 Coll., 515/2008 z. of., 117/2010 Coll., 145/2010 Coll., 408/2011 Coll., 180/2013 Coll., 207/2013 Coll., 311/2013 Coll., 506/2013 Coll., 35/2014 Coll., 198/2014 Coll., 314/2014 Coll., 324/2014 Coll., 91/2016 Coll. 125/2016 Coll.) and the implementing rules specify

- That the purpose of the act is, inter alia, preserving the diversity of conditions and life forms on Earth, as well as achieving and maintaining ecological stability of TSES and its whole-territory structure (biocentres, biocorridors, interactive elements).
- That the establishing and maintaining of TSES is a public interest.
- New "networks": the national network of protected areas and a European network of protected areas NATURA 2000.

(d) Act on Environmental Impact Assessment **No. *127/1994*** Coll. and subsequent ***24/2006*** Coll. defines TSES as an obligatory object of impact assessment.

(e) ***Water Act No. 364/2004 Coll.*** as amended and the Slovak National Council Act No. 372/1990 Coll. on offences (Water Act—based on Water Framework Directive 2000/60 of EP and EC) as amended (in particular No. 587/2004 Coll., 230/2005 Coll., 479/2005 Coll., 532/2005 Coll., 359/2007 Coll., 514/2008 Coll., 515/2008 Coll., 384/2009 Coll., 134/2010 Coll., 556/2010 Coll., 258/2011 Coll., 408/2011 Coll., 306/2012 Coll., 321/2012 Coll. (indirectly), 180/2013 Coll., 35/2014 Coll., 409/2014 Coll., 262/2015 Coll.), which defines

- The water plan of the Slovak Republic, river basin management plans which explicitly require coordination of water management tasks with the development and utilisation of the water protecting function of TSES.

(f) ***Act 7/2010 Coll.*** on flood protection (effective from February 1, 2010) as amended (as No. 80/2013 Coll., 71/2015 Coll.) enacts:

- That the long-term management plan of watersheds should project also the TSES, important landscape elements and the ecostabilising measures.
- These proposals within the plan of the food risk management shall be accepted as obligatory regulation in territorial plans,
- The TSES elements and other important landscape elements projected shall be accepted as corporate facility as stipulated by the Act on Land Arrangement and Land Ownership No. 331/1991 as amended.

(g) ***Act 326/2005 Coll.*** on forests as amended (in particular 275/2007 Coll., 359/2007 Coll., 360/2007 Coll., 540/2008 Coll., 499/2009 Coll., 117/2010 Coll., 96/2012 Coll., 345/2012 Coll., 115/2013 Coll., 180/2013 Coll., 182/2014 Coll., 125/2016 Coll.) determines that forest management plans shall promote maintaining ecological stability as one of the principal objectives of forest management.

Implementation of TSES into the laws governing spatial planning processes is a key element of the integrated landscape management.

The full application of the concept requires adopting many follow-up measures in legislation as well as the administration of state environmental protection.

References

Agger P, Brandt J (1988) Dynamics of small biotopes in Danish agricultural landscapes. Landscape Ecol 1(4):227–240

Ahern JF (2002) Greenways as strategic landscape planning: theory and application, Wageningen, Ill. Fig., 156p, Wageningen University

Antrop M (2013) A brief history of landscape research. In: Howard P, Ian Thompson I, Waterton E (eds) The Routledge companion to landscape studies. Routledge, Oxon, New York, pp 12–22

Bailey RG (1995) Description of the ecoregions of the United States. 2nd edn. rev. and expanded. Misc. Publ. No. 1931 (rev.), Washington D.C. USDA Forest Service, 108p

Barančoková M, Krajčí J, Kollár J, Belčáková I (2010) Landscape ecology—methods, applications and interdisciplinary approach. Institute of Landscape Ecology Slovak Academy of Sciences, Bratislava
Bastian O, Grunewald K, Khoroshev AV (2015) The significance of geosystem and landscape concepts for the assessment of ecosystem services: exemplified in a case study in Russia. Landscape Ecology 30(7):1145–1164
Bennett G (2002) Guidelines on the application of existing international instruments in developing the Pan-European ecological network. Nature and environment, No 124, Council of Europe Publishing, 101p
Bennett G, Mulongoy KJ (2006) Review of experience with ecological networks, corridors and buffer zones. Secretariat of the Convention on Biological Diversity, Montreal, Technical Series No 23, 100p
Bezák P, Izakovičová Z, Miklós L et al (2010) Reprezentatívne typy krajiny Slovenska. Bratislava, Ústav krajinnej ekológie SAV, 180p
Bonnin M (2007) The Pan-European ecological network–taking stock. Nature and environment, No 146, Council of Europe Publishing, 116p
Brandt J (1985) Small biotope structures as a synthesizing feature in agricultural landscapes. Proc IGU. Dessau: 52–61
Brandt J (1995) Ecological networks in Denmark. In: van Buuren M, Jongman R (eds) Ecological networks in Europe. Special issue of the journal Landschap, Wageningen, pp 63–76
Buček A, Lacina J, Löw J (1984) Teoretické východiská a typológia ÚSES. Pracovné materiály KRB, Agroprojekt Brno, p 12
Buček A, Lacina J (1984) Biogeografický přístup k vytváření územních systému ekologické stability krajiny. Zprávy Geografického ústavu ČSAV Brno 21(4):27–35
Buček A, Lacina, J (1995) Přírodovědná východiska ÚSES. In: Löw J et al Rukověť projektanta místního ÚSES, Doplněk Brno, pp 9–28
Buček A, Lacina J, Michal I (1996) An ecological network in the Czech Republik. Veronica, Special 11th issue
Bunce RGH, Watkins JW, Brignall P, Orr J (1996) A comparison of the environmental variability within the European Union. In: Jongman RHG (ed) Ecological and landscape consequences of land use change in Europe. European Centre for Nature Conservation, Publication Series on Man and Nature, 2, Tilburg
Chorley RJ, Kennedy BA (1971) Physical geography—a system approach. Prentice Hall International Inc, London, p 370
Cook EA, van Lier HN (eds) (1994) Landscape planning and ecological networks. Elsevier, New York, p 354
Csorba P (1989) Tájstabilitás és ökogeográfiai stabilitás. Földrajzi Értesítő 38(3–4):395–410
Csorba P (1996) Landscape ecological corridors on the East Foothills of the Tokaj Mountains (Hungary). Arbeiten aus dem Institut für Landschaftsökologie, WWU Münster 2:217–229
Csorba P (2008) Landscape ecological fragmentation of the small landscape units (Microregions) of Hungary based on the settlement network and traffic infrastructure. Ekológia (Bratislava) 27(1):99–116
de Groot RS, Alkemade R, Braat L, Hein L, Willemen L (2010) Challenges in integrating the concept of ecosystem services and values in landscape planning, management and decision making. Ecological Complexity 7(3):260–272
Drdoš J, Miklós L, Kozová, M, Urbánek J (1995) Základy krajinného plánovania. Učebné texty. Technická univerzita vo Zvolene, 173p
Ellenberg H (ed) (1973a) Ökosystemforschung. Springer, Berlin
Ellenberg H (1973b) Versuch einer Klassifikation der Ökosysteme nach funktionalen Gesichtspunkten. In: Ellenberg H (ed) Ökosystemforschung. Springer, Berlin, pp 235–265
Fabos JG (1996) Greenways: the beginning of an international movement. Elsiever, Amsterdam

Forman RTT (1990) Ecologically sustainable landscapes: the role of spatial configuration. In: Zonneveld IS et al (eds) Changing landscapes: an ecological perspective. Springer, New York, pp 261–278
Forman RTT, Godron, M (1981) Patches and Structural Components for a Landscape Ecology. BioScience 31(10):733–740
Forman RTT, Godron M (1986) Landscape ecology. Wiley, New York
Forman RTT, Godron M (1993) Krajinná ekologie. Academia, Praha
Grodzinsky MD (1987) Stability of geosystems: theoretical approach to the analysis and methods of quantitative evaluation. Proc Acad Sci USSR, Geogr Ser 6:5–15
Grodzinski MD (2005) Piznannya landschaftu: misce i prostir (Understanding Landscape. Place and Space. In Ukrain). Kiev. U 2-x t. Vidavnicshopoligrafichesky centr Kiivsky universitet. T.1
Hanski I (1998) Metapopulation dynamics. Nature 396(6706):41–49
Hellmund PC, Smith D (2013) Designing greenways: Sustainable landscapes for nature and people, Island Press
Hilty AJ, Lidicker VZ Jr, Merenlender A (2006) Corridor ecology: the science and practice of linking landscapes for biodiversity conservation. Island press, Washington, p 344
Hreško J, Mederly P, Petrovič F (2003) Landscape ecological research with support of GIS tools in prepration of landscape-ecological plan (model area of the Považská Bystrica city). Ekológia (Bratislava) 22(2):195–212
Húsenicová J, Ružičková J, Klinda J, Miklós L et al (1992) Generel nadregionálneho územného systému ekologickej stability Slovenskej republiky. Slovenská komisia ŽP, Bratislava
Isachenko AG (1981) Predstavlenije o geosisteme v sovremennoj fizičeskoj geografiji. Izv. VGO Leningrad 113(4):297–306
Izakovičová Z, Hrnčiarová T, Králik J, Liška M, Miklós L, Moyzeová M, Pauditšová E, Ružičková H, Šíbl J, Tremboš P (2000) Metodické pokyny na vypracovanie projektov regionálnych územných systémov ekologickej stability a miestnych územných systémov ekologickej stability. MŽP SR, Združenie KRAJINA 21, Bratislava, 155p
Izakovičová Z, Miklós L, Moyzeová M et al (2011) Model reprezentatívnych geoekosystémov na regionálnej úrovni. Ústav krajinnej ekológie SAV, Bratislava, p 88
Jančura P et al (1994) Manuál k metodike ÚSES, I.—IV., Ministerstvo životného prostredia SR, Bratislava
Jongman RHG (1995) Nature conservation planning in Europe: developing ecological networks. Landscape Urban Plan 32:169–183
Jongman RHG (1996) Research priorities: scientific concepts and criteria. In: Perspectives on ecological networks. Eur Centre Nat Conserv Ser Man Nat 1:151–160
Jongman RHG (2002) Homogenisation and fragmentation of the European landscape: ecological consequences and solutions. Landscape Urban Plan 58(2–4):211–221
Jongman RHG (2008) Ecological networks, from concept to implementation. In: Hong SK, Nakagoshi N, Morimoto Y (eds) Landscape ecological applications in man-influenced area. Springer, Berlin, pp 57–69
Jongman RHG, Kristiansen IB (2001) National and regional approaches for ecological networks in Europe. Nature and environment, No. 110, Council of Europe Publishing, 86p
Jongman RHG, Külvikb M, Kristiansenc IB (2004) European ecological networks and greenways. Landscape Urban Plan 68(2–3):305–319
Jongman RHG, Pungetti G (2004) Ecological networks and greenways: concept, design, implementation. United Kingdom at the University Press, Cambridge, p 153
Jongman RHG, Bouwm IM, Griffioen A, Jones-Walters L, van Doorn AM (2011) The Pan European ecological network: PEEN. Landscape Ecol 26(3):311–326
Kiss Á, Szabó M (2016) Zostane Zem vhodná pre ľudí? In: Mezřický, J., Mezřický, V. (eds) Překročit práh. (snad) náměty k úvahám. Jiří a Václav Mezřický - BOOM TISK, Kolín, pp 159–178. ISBN: 978-80-972396-3-3

Konkolyné Gyuró É (2005) The Role of Green Corridors in Landscape Restoration. In: Konkoly Gyuró É (ed) Greenways. Conference Presentations on Ecological Corridors, Green Corridors, Sopron, Hungary, pp 1–21

Kočická E, Kočický D (2014) Databáza abiotických komplexov ako krajinno-ekologický podklad pre integrovaný manažment krajiny v Slovenskej republike. Acta Facultatis Ecologiae (Zvolen), TU Zvolen, 31:35–50

Kozová M, Hrnčiarová T, Drdoš J et al (2007) Landscape Ecology in Slovakia. Development, Current State and Perspectives. Monograph. Contribution of the Slovak Landscape Ecologists to the IALE World Congress 2007 and to the 25th Anniversary of IALE. Bratislava: Ministry of the Environment of the Slovak Republic, Slovak Association for Landscape Ecology–IALE-SK, 2007, CD ROM

Krcho J (1991) Georelief as a subsystem of landscape and the influence of morphometric parameters of georelief on spatial diferentation of landscape-ecological processes. Ecology (CSFR), Bratislava 10(2):115–158

Kubeš J (1996) Biocentres and corridors in a cultural landscape. A critical assessment of the territorial system of ecological stability. Landscape Urban Plan 35(4):231–240

Kuttner M, Hainz-renetzeder CH, Hermann A, Wrbka T (2013) Borders without barriers–Structural functionality and green infrastructure in the Austrian-Hungarian transboundary region of Lake Neusiedl. Ecol Indic 31:59–72. https://doi.org/10.1016/j.ecolind.2012.04.014

Macarthur RH, Wilson ED (1967) The theory of island biogeography. Princeton University Press, Princeton, p 233

Mander U, Jagomagi J, Kiilvik M (1987) Network of compensative areas as an ecological infrastructure of territories. In Connectivity in Landscape Ecology. Edited by K.-F. Schreiber. Proc. 2nd Internat. Seminar of the International Association for Landscape Ecology. Miinstersche Geograph. Arbeiten 29:35–38

Míchal I (1992) Ekologická stabilita. Veronica, Brno, p 244

Míchal I et al (1991) Územní zabezpečování ekologické stability: Teorie a praxe. Praha, MŽP ČR, p 150

Miklós L (1991) Zásady slovenskej ekologickej politiky. Životné prostredie 25(4):174–178

Miklós L, Diviaková A, Izakovičová Z (2011) Ekologické siete a územný systém ekologickej stability. Vydavateľstvo TU vo Zvolene

Miklós L, Izakovičová Z (1997) Krajina ako geosystém. VEDA, SAV, Bratislava, p 152

Miklós L et al (1985) Ekologický generel ČSSR. Časť SSR. I. etapa: Priestorová diferenciácia územia z ekologického hľadiska. Záverečná správa P16-121-402/01. ČSŽP Bratislava. ÚEBE CBEV SAV Bratislava, Stavoprojekt Banská Bystrica, 152p

Miklós L, Kozová M, Ružička M et al (1986) Ekologický plán využívania Východoslovenskej nížiny v mierke 1:25.000 (Ecological management plan of the Eastern lowlands at a scale of 1:25,000, in Slovakian). In: Ekologická optimalizácia využívania VSN, 3:5–312. ÚEBE CVEB SAV. Bratislava

Miklós L, Izakovičová Z, Boltižiar M, Diviaková A, Grotkovská L, Hrnčiarová T, Imrichová Z, Izakovičová Z, Kočická E, Kočický D, Kenderessy P, Miklós L, Mojses M, Moyzeová M, Petrovič F, Špinerová A, Špulerová J, Štefunková D, Valkovcová Z, Zvara I, Hreško J, Lauko V, David S (2006) Atlas of representative geoecosystems of Slovakia. Ústav krajinnej ekológie SAV, Bratislava, 123p

Miklós L, Hrnčiarová T (eds) et al (2002a) Atlas krajiny Slovenskej republiky. 1. vydanie, MŽP SR Bratislava, SAŽP Banská Bystrica, 344p

Miklós L, Kočická E, Kočický D (2002b) Nadregionálny územný system ekologickej stability. In: Miklós L, Hrnčiarová T (eds) Atlas krajiny SR, MŽP SR, SAŽP Banská Bystrica, 1. vydanie, pp 258–259

Miklós L, Kočická E, Kočický D, Diviaková A (2015) Geosystémy ako krajinnoekologická základňa pre integrovaný manažment krajiny. (Geosystems as a landscape-ecological basis for integrated landscape management—in Slovak). Technická univerzita Zvolen

Miklós L, Špinerová A, Belčáková I, Diviaková A (2018) Instruments and tools for sustainable land use and spatial development processes. In: Belčáková I, Miklós L (eds) Landscape considerations in spatial planning processes. Peter Lang: Frankfurt a.M. et al., ISBN 978-3-631-74971-5
Miklós M, Krcho J, Hrnčiarová T, Matečný I, Kozová M (1997) Morfometrické ukazovatele reliéfu a ich účelové interpretácie pre plánovacie procesy. Nadácia Katedry UNESCO pre ekologické vedomie, Banská Štiavnica, p 94
Naveh Z, Lieberman AS (1993) Landscape ecology-theory and application, 2nd edn. Springer, New York
Návrh národnej ekologickej siete Slovenska–NECONET (1996) Nadácia IUCN, Svetová únia ochrany prírody, Bratislava
Neef E (1963) Topologische und chorologische Arbeitsweisen in der Landschaftsforschung. PGM 107(4):249–259
Neef E (1967) Die theoretischen Grundlagen der Landschaftslehre. H. Haack, Gotha, Leipzig
Moyzeová M (2010) Navrhovanie územných systémov ekologickej stability na vybraných modelových územiach. In: Životné prostredie. 44(3):138–142
Moyzeová M, Grotkovská L (2006) Environmental aspects of wetland evaluation (model area of the Paríž water stream). In: Ekológia (Bratislava). 25(1):169–178
Odum EP (1975) Ecology: the link between the natural and the social sciences, 2nd edn. Holt Rinehart and Winston, London, p 244
Opdam P, Foppen R, Reijnen R, Schotman A (1995) The landscape ecological approach in bird conservation: integrating the metapopulation concept into spatial planning. Ibis 137:139–146
Pan-European Biological and Landscape Diversity Strategy (1996) Council of Europe Press, Strassbourg
Preobrazhensky VS, Minc AA (1973) Sootnoshenye ponyaty geosystema a ekosystema. In: Práce a materiály z biológie krajiny 20. Proceedings of 3rd International Symposium on the Landscape Ecological Research. ÚBK SAV, Bratislava
Renetzeder C, Wrbka T, Mücher S;, Eupen M V, Kiers M (2010a) Landscape Structure Analysis in selected European Regions and its use for Sustainability Impact Assessment. In: Andel J, Bicik I, Dostal P, Lipsky Z, Shahneshin SG (eds) Landscape Modelling. Geographical Space, Transformation and Future Scenarios. Series: Urban and Landscape Perspectives, vol 8, 236p
Renetzeder CH, Schindler S, Jpeterseil J, Prinz MA, Mucher S, Wrbka T (2010b) Can we measure ecological sustainability? Landscape pattern as an indicator for naturalness and land use intensity at regional, national and European level. Ecol Ind Elsevier 10(1):39–48
Richling A, Solon J (1993) Ekologia krajobrazu. Wydawnictwo Naukowe PWN, Warszawa
Rientjes S, Drucker G (1996) An Introduction to the Pan-European biological and landscape strategy—background, philosophy and summary, The Pan-European Biological and Landscape Strategy. Diversity Strategy. Council of Europe, UNEP & ECNC, 10-14. 32
Ružička M, Miklós L (1982) Landscape ecological planning (LANDEP) in the process of territorial planning. Ekologia (CSSR) 1(3):297–312
Ružička M, Miklós L (1990) Basic Premises and Methods in Landscape Ecological Planning and Optimization. In: Zonneveld IS, Forman RTT (eds) Changing landscapes: an ecological perspective. Springer, New York, pp 233–260
Ružička M et al (1982a) Ekologické hodnotenie predpokladov na rozvoj poľnohospodárstva v záujmovom území vodárenskej nádrže Klenovec. Životné prostredie, Bratislava 16(6):311–317
Ružičková H, Ružička M (1973) Druhotná štruktúra krajiny ako kritérium biologickej rovno¬váhy. Quaestiones geobiologicae 12:23–62
Ružička M, Jurko A, Kozova M, Zigrai F, Svetlosanov V (1982b) Evaluation methods of landscape stability on agricultural territories in Slovakia. VIth International symposium on problems of landscape ecological research. Oct 25–30, 1982, Piešťany (CSSR)
Schilleci F, Todaro V, Lotta F (2017) Connected lands: new perspectives on ecological networks planning. Springer, Berlin, 85p

Schreiber KF (ed) (1988). Connectivity in landscape ecology. Proceedings of the 2nd International Seminar of the IALE in Münster. Münsterische Geographische Arbeiten, 29, Schöningh, Paderborn, Germany, 255p
Sepp K, Kaasik A (eds) (2002) Development of national ecological networks in the Baltic countries in the framework of the Pan-European ecological network. IUCN, Pracownia Poligraficzna ARWIL s.c., Warszawa, 183p
Sochava VB (1977) Vvedenje v učenije o geosystemach Nauka, Novosibirsk
Solon J (1988) The spatial distribution of vegetation units as a result of habitat and synanthropization pattern. VIIIth International Symposium on Problems of Landscape Ecological Research. Oct 3–7, 1958, Zemplinska sirava (CSSR)
Špinerová A (2015) Štruktúra krajiny ako regulátor dynamiky pohybu vody a materiálu. Zvolen, Vydavateľstvo TU vo Zvolene
Špulerová J, Dobrovodská M, Izakovičová Z, Kenderessy P, Štefunková D, Petrovič F (2013) Developing a strategy for the protection of traditional agricultural landscapes based on a complex landscape-ecological evaluation (the case of a mountain landscape in Slovakia). Moravian Geographical Reports 21(4):15–26
Štefunková D (2011) Vybrané výsledky výskumu starej vinohradníckej krajiny Svätého Jura. In: Ekologické štúdie: Vedecký časopis venovaný aktuálnym problémom ekológie, krajinnej ekológie a príbuzných vedných disciplín, 2(1-2):126–131
Štefunková D, Hanušin J (2015) Analysis of the spatial and temporal distribution of selected landscape diversity indexes in detailed scale (Example of viticultural landscape Svätý Jur). In: Halada L', Bača A, Boltižiar M (eds), 2015, Landscape and Landscape Ecology. Proceedings of the 17th International Symposium on Landscape Ecology. Institute of Landscape Ecology, Slovak Academy of Sciences, Bratislava, p. 185–191
Szabó M (2007) Tájszerkezeti változások a Felső-Szigetközben az elmúlt 20 évben. Földrajzi Közlemények. CXXXI. (LV) 1(2):55–74
Szabó M (2008) A bioszféra sokfélesége – az élővilág változatossága és veszélyeztetettsége. Földrajzi Közlemények 132(4):441–449
Tansley AG (1935) The use and abuse of vegetational concepts and terms. Ecology 16(3):284–307
Tardy J, Végh M (eds) (2006) National Ecological Networks of European Countries 1:4,000,000, European Centre for Nature Conservation
Tillmann JE (2005) Habitat Fragmentation and Ecological Networks in Europe. GAIA. Ecol Perspect Sci Soc 14(2):119–123
von Bertalanffy L (1968) General system theory. Foundations, development and applications. George Brazileer, New York. Penguin Books
Wrbka T, Stocker-Kiss A, Schmitzberger I, Peterseil J (2005) Network Landscapes of Austria – Identification and Ecological Assessment. In: Konkoly GÉ (ed) Greenways. Conference Presentations on Ecological Corridors, Green Corridors, Sopron, Hungary, pp 47–62

Chapter 3
Methodical Base of Development of the Territorial System of Ecological Stability

Abstract The definition of TSES is as follows: TSES is a whole-territory covering system of an ecologically optimum structure of the landscape composed of elements with different degrees of ecological stability and different uses, but in its entirety ensuring both the ***internal functioning*** of individual core ecosystems and the functionality of the ***spatial relations*** among them as a precondition for the maintenance of the spatial landscape-ecological stability and so the diversity of both conditions and forms of life. TSES has two equally important essential parts: the skeleton of TSES—a system of **biocentres, biocorridors and interactive elements**, and a system of **ecostabilising measures**. Both these parts are of the same importance. The functioning of TSES can be maintained only by safeguarding both of them. The criteria for determining the functions of landscape elements for the TSES are following:

- Selective criteria—determine whether a landscape element is able to fulfil the functions as shelter, nourishment and reproduction. An important criterion is geoecological representativeness. These are the strategic criteria mainly for selection of biocentres.
- Localisation criteria—determine whether the landscape elements perform a function within the spatial arrangement of the biocentres, biocorridors and interactive elements as the change of genetic information and migration, as well as specific functions as retention, anti-erosive, microclimatic, hygienic, aesthetic functions, etc.
- Realisation criteria—determine the conditions for the realisation of the TSES in relation to the human activities. Most important indicator in this group is the spatial arrangement of the land-use elements and the legal frame of nature and nature resources protection.

TSES in Slovakia is projected on three hierarchical levels: the General Plan of the supra-regional TSES related to the whole territory of Slovakia, the regional TSES-s related to the districts of Slovakia and the local TSES-s related to the municipalities. TSES is incorporated in several acts. For an effective implementation of TSES, the decisive importance has the legal determination of the position of the TSES in the territorial planning documentations where the elements of the TSES are defined as obligatory regulations on all level of the planning process.

L. Miklós et al., *Ecological Networks and Territorial Systems of Ecological Stability*, https://doi.org/10.1007/978-3-319-94018-2_3

Keywords Biocentres · Biocorridors · Interactive elements · Ecostabilising measures · Criteria · Hierarchy

After the gradual development of its methodology, the TSES can be defined as follows (Miklós et al. 2018):

The territorial system of ecological stability (TSES) is a whole-territory covering system of an ecologically optimum structure of the landscape composed of individual elements with different degrees of ecological stability and different uses, but in its entirety ensuring: ***internal functioning*** of individual core ecosystems, ***functionality*** of the spatial relations among them as a precondition for the maintenance of the spatial landscape-ecological stability and so the diversity of both conditions and forms of life.

The definition included in the Act No. 543/2002 on nature and landscape protection reads as follows: "the Territorial System of Ecological Stability is a whole-space system of interconnected ecosystems, their components and elements ensuring the diversity of conditions and forms of life in the landscape. The system consists of biocentres, biocorridors and interactive elements of over regional, regional or local level (§2 of the Act No. 543/2002 Coll.)".

The methodical base of TSES was described also in many publications, as Izakovičová et al. (2000), Miklós (1996), Miklós et al. (2006, 2011, 2018). In this publication, we present the methodical base in the form as described in the last mentioned publication:

TSES has two equally important essential parts:

- The skeleton of TSES—a system of biocentres, biocorridors and interactive elements, and
- A system of ecostabilising measures.

Both these parts are of the same importance, and functioning of TSES can be maintained only by safeguarding both of them.

Numerous TSES projects have been elaborated based on these theoretical and methodological principles. Nevertheless, there is a permanent need for practical improvement of the practical procedures. The characteristics of the elements of the procedure are described in following chapters.

3.1 Elements of the Frame/Skeleton of Territorial System of Ecological Stability and Their Functions

The elements of TSES are represented by various types of geoecosystems. To maintain the desired state of the system, it is crucial to sustain adequate ecological relations between them. The key elements of the skeleton of TSES—biocentres, biocorridors and interactive elements—need to ensure proper functioning of the following critical relationships (Miklós et al. 2018):

(a) **Biocentres** maintain the food chain as well as the conditions for a reproductive cycle (reproduction and raising offspring) and preservation and provide protection, resting place and shelter. Míchal et al. (1991) define a biocentre as "a landscape segment, whose size and ecological conditions allow long-term (preferably permanent) existence of species or communities of native fauna and flora and their genetic resources". Act No. 543/2002 defines a biocentre as an ecosystem or a group of ecosystems, which create sustained conditions for reproduction, shelter and nutrition of living organisms, and the preservation and natural development of their indigenous communities.
(b) **Biocorridors and interactive elements** facilitate overcoming barriers that isolate ecosystems and maintain the exchange of genetic information and migration as well as the interaction between various ecosystems of different degree of ecological stability (particularly important are linear communities and ecotones).

Ecosystems with a high degree of ecological stability (forests, landscape greenery, grasslands) perform different practical ecological functions such as soil and water protection, microclimatic, hygienic, aesthetic and other functions.

Míchal et al. (1991) define biocorridor as "Landscape segment which interconnects biocentres and thus allows migration of organisms although they might not necessarily provide permanent conditions of existence". According to the authors, the term "migration" also encompasses the exchange of genetic information.

According to the Act No. 543/2002 Coll., a biocorridor is a spatially interconnected set of ecosystems that connects biocentres, allows the migration and exchange of genetic information of living organisms and its communities, and is linked with interactive elements.

In reality, however, biocorridors harbouring numerous species (especially the more mobile ones) are often spatially disconnected. For example, in our conditions, the biocorridors of xerothermic species may consist of a chain of more or less spatially unlinked (isolated) lower-level biocentres at the regional and higher level. The same view has been expressed by the authors of the European Ecological Network (EECONET) (Bennett 1991).

Ružičková (2002) recommends deleting the explanatory note of interactive elements, since it does not reflect the reality as interactive elements may or may not be linked with biocorridors.

Míchal et al. (1991) define an interactive element as "landscape segment, usually of an ecotone character, mediating favourable effects of other ecologically important landscape segments (biocentres and biocorridors) commonly at the local level".

An interactive element is a specific ecosystem, its elements or a group of ecosystems (e.g. permanent grassland, bog, forest, lake) which are interconnected with biocentres and biocorridors and which positively affect the surrounding parts of the landscape changed or disturbed by human activities (Act no. 543/2002 Coll.).

However, Ružičková (2002) argues that a lake, bog or forest may also function as a biocentre, so the presented example does not apply. Furthermore, she complements the definition as follows: "it includes mostly small-area landscape elements such as escape coverts or groups of trees or narrow linear stands such as treelines, windbreaks, riparian vegetation and the like".

Šíbl (1994) also points out an imperfection stating that Act No. 543/2002 Coll. does not specify whether the interconnection between an interactive element and biocentres or biocorridors is spatial or only functional and how to interpret the term "ecosystem element".

Löw et al. (1995) classify biocentres and biocorridors by their functionality, formation and development, representativeness, habitat diversity, diversity of current biocenosis (real vegetation), formation type, geoecological (horizontal relations), biogeographical location; in addition, biocorridors are categorised by connectivity and similarity of linked biocentres.

3.2 Criteria for Determination of Territorial System of Ecological Stability Functions

The following are the criteria for determining the functions of landscape elements of the TSES:

(a) **Selective** criteria—determine whether the landscape element in question is suitable to perform required ecological functions. An important criterion is a landscape-ecological significance and geoecological representativeness, i.e. whether a landscape element is important for maintaining diversity of conditions and life forms in a territorial unit, selected group of communities, populations or species. These strategic criteria are employed to determine the landscape elements and types that need to be integrated into TSES as biocentres.
They are determined primarily based on the indicators of biotic features of the current landscape structure (Miklós and Izakovičová 1997).

(b) **Location** criteria—determine functionality of the landscape structures in the spatial arrangement of TSES, especially from the point of the suitability of biocentre location, as well as functions of biocorridors and interactive elements. Another special location criterion evaluates whether a concerned landscape element performs its specific functions (soil and water protection, erosion prevention filtration, microclimatic, hygienic, aesthetic).
These functions are determined by the property indicators of the abiotic features of landscape elements (Miklós and Izakovičová 1997).

(c) **Realisation** criteria—determine the possibility of TSES implementation with respect to the character of human activity in the area. They evaluate if the position of a TSES element is feasible or not in the area, what necessary measures need to be introduced, and they also identify the most critical locations within TSES. The main indicator in this group is the spatial arrangement of the current landscape structure, location and proportion of areas of natural, semi-natural and artificial landscape elements, barrier effect of landscape elements as well as

the occurrence of socio-economic landscape elements (Miklós and Izakovičová 1997).
From the practical perspective, the current legislative protection is of extra importance.

Legally protected areas and landscape elements form the skeleton of TSES.

The establishment of TSES is a complex process and all—the selective, location and realisation criteria must be taken into account.

Developing of TSES also requires a *proposal for* "ecostabilising measures", and application of the methodological principles utilised in landscape-ecological planning (LANDEP) (Ružička and Miklós 1982, 1990).

3.3 Hierarchical Structure of the Territorial System of Ecological Stability

Like every phenomenon in space, TSES should be also interpreted on different hierarchical levels. In Slovakia, TSES is organised on three levels: ***supra-regional, regional and local***. The hierarchical levels are important mostly for developing TSES projects and their implementation as well as the integration of TSES into spatial plan at different levels.

The logic of the hierarchy of environmental conditions implies that a higher TSES provides a framework for a lower, i.e. the principles applied in the supra-regional TSES cannot be negated by regional interests and regional by local interest.

TSES will fulfil its function only after each element has been fully implemented at the local level.
That is the real TSES.
There is no supra-regional or regional tree, bog, water areas; there are just local landscape elements.

Act No. 543/2002 Coll. defines TSES as an integrated structure of interconnected ecosystems, which ensures diversity of conditions and life forms in the landscape. This system is represented by biocentres, biocorridors and interactive elements at different hierarchical levels, as supra-regional—biospheric and provincial, regional and local. However, it needs to be emphasised that **the system of measures to ensure ecologically optimum organisation and land use—ecostabilising measures**—is an important part of the implementation of the whole-territory TSES mainly on regional and local level. The skeleton of TSES (biocentres and biocorridors) as well as the system of ecostabilising measures are both essential and equally important to safeguard proper functionality of TSES.

According to Decree of the Ministry of Environment, 24/2003 (§ 22, Section 1 and 2) **General Plan supra-regional territorial system of ecological stability of the Slovak Republic** (hereinafter as "the General Plan" or GTSES) is the basis for the strategies to preserve the diversity of conditions and life forms in the country. The General Plan reflects the current state of the ecological quality of the entire territory of the state and includes basis for achieving ecological stability. The General Plan is a foundation for the preparation of documents of the regional and local TSES. It was elaborated in scale 1:500,000 or 1:200,000 (Fig. 3.1).

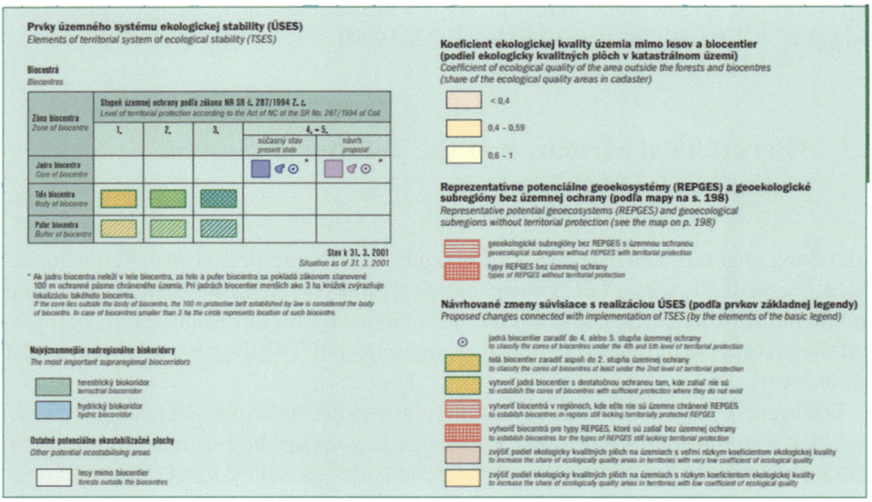

The legend to the Fig. 3.1

The documents of the **regional territorial system of ecological stability** (hereinafter as RTSES) are focused on the protection of diversity of conditions and life forms and on achieving ecological stability usually at the level of the district; it is a basis for the territorial plan of the region and for the territorial municipality plan, for preparing documents for the local territorial system of ecological stability, for decision-making of nature protection bodies and for practical management of specially protected parts of nature and landscape (MoE Decree No. 24/2003, § 22, Section 3-6).

The document of the regional territorial system of ecological stability contains:

(a) Territorial projection of nature and landscape protection with an overview of selected parts of nature and landscape, which are of particular importance for preservation of biological diversity and ecological stability of the landscape regardless of their protection level (hereinafter as "specially significant parts of nature and landscape"),
(b) Territorial projection of load on nature and landscape with assessment of a load degree and changes of landscape triggered by human activity,

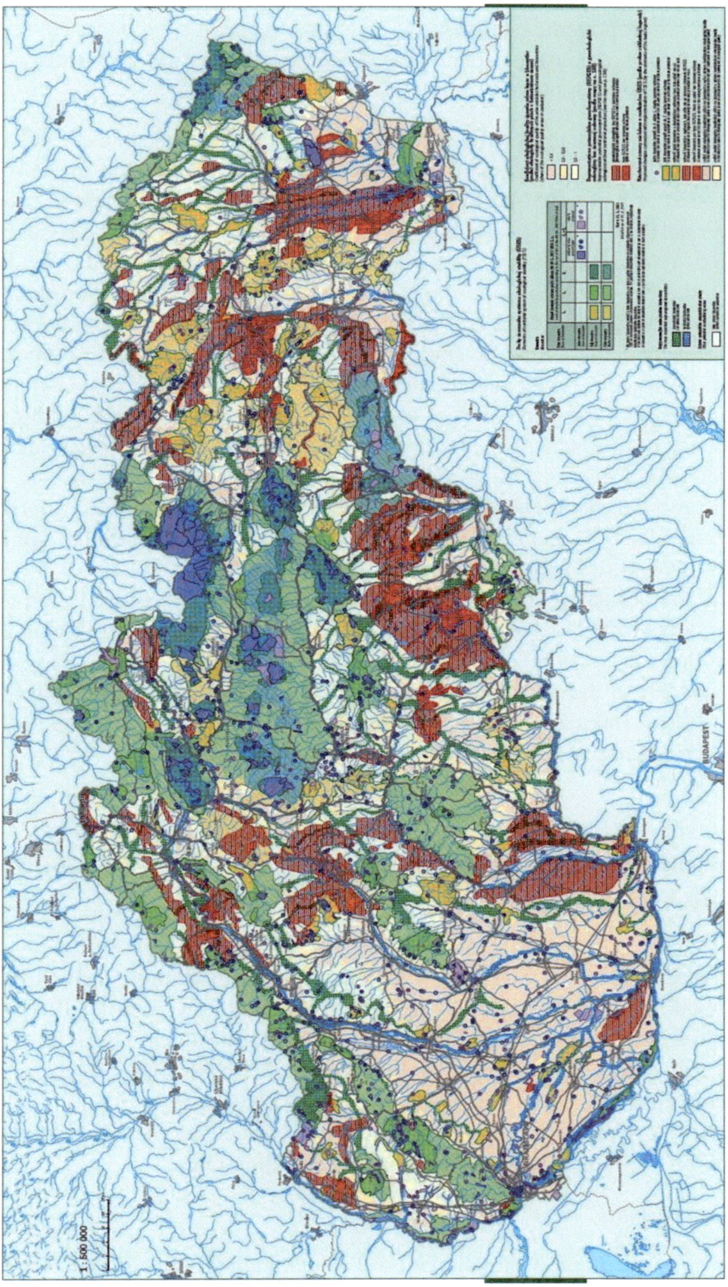

Fig. 3.1 General Plan of the territorial system of ecological stability of SR

(c) Territorial projection of ecological stability of landscape with categorisation by the proportion of specially significant parts of nature and landscape and degree of their anthropogenic changes,
(d) Proposal of the measures specifying regulations for nature and landscape protection and development.

The document of the regional territorial system of ecological stability (RTSES) is elaborated usually for a territory of a district, apart from the territory of a protected landscape area or a national park, including its protection zone, where the function of the regional territorial system of ecological stability is fulfilled by documents as specified by § 21 Section (1) of MoE Decree No. 24/2003. Thus, the RTSES constitutes the fundamental landscape-ecological basis for the elaboration of the land-use planning documentation of a higher regional unit, and for other departments and their relevant documentations (nature conservation, territorial development, land consolidation projects, forest management plans, river basins management plans a.o).

The map scale of the document of the RTSES usually corresponds with the map scale of the territorial plans; most often are represented on maps on a scale 1:50,000. For example, see Appendix B.

The document of the **local territorial system of ecological stability** (hereinafter referred to as MUSES) is a basic document of protection of diversity of conditions and life forms, and for achieving ecological stability at the local level.

It forms the obligatory basic material for the municipal territorial plan and for territorial plan of municipal zones, as well as for consolidation projects (MoE Decree No. 24/2003, § 22, Section 7-8).

The map scale of the document of the R TSES usually corresponds with the map scale of the territorial plans; most often are represented on maps on a scale 1:10,000 or 1:5000. For example, see Appendix B.

3.4 Position of the Territorial System of Ecological Stability in Territorial Planning

Territorial planning is one of the most significant legal institutions, which, through its conceptual and operational tools, decisively affects the way the land is utilised, while guaranteeing preservation of its natural and cultural values. This is emphasised in the first part, the first section and the first article of Act No. 50/1976 on Territorial Planning and Building Order (the Building Act) as amended.

The territorial system of ecological stability is mentioned in the following sections of Act No. 50/1976 on Land-use Planning and Building Order:

§ 7a Other materials

(1) In the land-use planning activity, the existing documents and files of information that include information on land are used (hereinafter as “other materials”).

(2) From the documents referred to in paragraph 1, there are obligatorily used:

(a) Strategies of sustainable development, strategy of state environmental policy, environmental action plans and other overall plans,
(b) Projects of land arrangement, forests, water-management, land irrigation and melioration,
(c) **Documents of the territorial system of ecological stability**, projects of protection of nature and landscape, nature and landscape care programmes,
(d) Programmes of protection of cultural and historical heritage,
(e) Programmes of waste management,
(f) Overall plans for development of individual areas of communal sphere and strategies and programmes of regional development.

§ 13 Binding part and guiding part: Proposal and regulations

(3) In the binding part of land-use planning documentation there are approved the following principles and regulatives:

(a) In the regions: spatial arrangement and functional use of region, **territorial system of ecological stability**, environmental care, creation of landscape, protection and economical use of natural resources, protection of cultural monuments, monumental reservations, monumental zones and important landscape elements, arrangement of public transport and technical facilities, establishment of areas for public buildings and for protective parts of landscape,
(b) In the municipality: spatial arrangement and functional use of the municipal area, boundaries of built-up area, arrangement of public transport, civil and technical facilities, establishment of areas for public buildings, for carrying out the sanitation and for protected parts of landscape, protection and use of natural resources, cultural-historical values and important landscape elements, **territorial system of ecological stability**, environmental care, creation of landscape including green areas.

§ 39b Land-use decision

(4) Decision on the land-use is required for:

(a) Performing land modifications which in essence alter the **territorial system of ecological stability**, the appearance of the landscape, use of important landscape elements and drainage conditions.

§ 139a Terms of land-use planning

(5) Ecologically optimal spatial arrangement and functional use of landscape (landscape-ecological plan) is a comprehensive process of mutual harmonising of spatial requirements of economic and other activities of man with the landscape-ecological conditions that result from the landscape structure. Ecologically optimal spatial arrangement and functional use of landscape at the

same time ensure the suitable stability of spatial structure of landscape, protection and rational use of nature, biodiversity and natural resources, creation and protection of **the territorial system of ecological stability** and immediate environment of man. Landscape structure and its elements are showed as limits, restrictions or supporting factors of required activities in the given territory.

(6) Territorial development is the development that in a sustainable way satisfies the fundamental living requirements of people in the territory while it does not decrease its diversity, it ensures the optimum spatial arrangement and functional use of land, environmental safety and suitability of buildings and facilities, creation and preservation of **the territorial system of ecological stability**, economical use of natural resources, protection of natural and cultural heritage.

3.5 Results of the Territorial System of Ecological Stability

The compiling process of the TSES results in a system of proposals, which can be further subdivided into:

(a) Proposals for the skeleton elements of TSES—**biocentres and biocorridors**. An essential component of these documents is a proposal to maintain or reinforce the present legal protection of the current landscape elements as well as proposals for establishment of new elements if the territory in question does not contain a representative geoecosystem suitable to be a **biocentre or biocorridor**.

(b) Proposals **to improve overall ecological quality of the current landscape structure**. This step includes the proposal for the modification of land use, especially its structure and location of non-forest and shrub vegetation as well as grasslands, which can be a part of biocorridors or interactive elements.

(c) Proposals for so-called **ecostabilising measures**. These ensure the whole-territory character of TSES. The proposals should specify the optimum use of the territory and agrotechnical, agrochemical and agroameliorative measures. The ecological optimisation of spatial landscape organisation is addressed by the methodology of **land**scape-**e**cological **p**lanning (**LANDEP**), which represents a systematically organised special-purpose complex of landscape-ecological methodologies and methods (Ružička and Miklós 1982), and thus the methodological procedures employed by LANDEP constitute a basis for elaboration of the TSES documents (MoE SR 1993; Izakovičová et al. 2000).

(d) Proposals for **necessary legislative measures** to ensure proper functionality of selected TSES elements.

TSES is complete only if it incorporates the proposals contained in all the aforementioned groups. It is a comprehensive and whole-space covering TSES concept.

References

Bennett G (ed) (1991) Towards a European ecological network. The Nederlands, Institute for European Environmental Policy, Arnhem, 80 pp

Izakovičová Z, Hrnčiarová T, Králik J, Liška M, Miklós L, Moyzeová M, Pauditšová E, Ružičková H, Šíbl J, Tremboš P (2000) Metodické pokyny na vypracovanie projektov regionálnych územných systémov ekologickej stability a miestnych územných systémov ekologickej stability. MŽP SR, Združenie KRAJINA 21, Bratislava, 155 pp

Löw J et al (1995) Manual for designers of the Local Territorial System of Ecological Stability. Methods for elaboration of the documentation. Appendix, Brno (in Czech), 124 pp

Míchal I et al (1991) Územní zabezpečování ekologické stability: Teorie a praxe. MŽP ČR, Praha, p 150

Miklós L (1996) The concept of the territorial system of ecological stability in Slovakia. In: Jongmann RHG (ed) Ecological and landscape consequences of land-use change in Europe. ECNC publication series on Man and Nature 2, Tilburg

Miklós L, Izakovičová Z (1997) Krajina ako geosystém. VEDA, SAV, Bratislava, 152 pp

Miklós L, Izakovičová Z, Boltižiar M, Diviaková A, Grotkovská L, Hrnčiarová T, Imrichová Z, Izakovičová Z, Kočická E, Kočický D, Kenderessy P, Mojses M, Moyzeová M, Petrovič F, Špinerová A, Špulerová J, Štefunková D, Valkovcová Z, Zvara I, Hreško J, Lauko V, David S (2006) Atlas of representative geoecosystems of Slovakia. ÚKE SAV, Bratislava

Miklós L, Diviaková A, Izakovičová Z (2011) Ekologické siete a územný systém ekologickej stability. Vydavatel'stvo TU vo Zvolene

Miklós L, Špinerová A, Belčáková I, Diviaková A (2018) Instruments and tools for sustainable land use and spatial development processes. In: Belčáková I, Miklós L (eds) Landscape considerations in spatial planning processes. Peter Lang, Frankfurt a.M. et al., ISBN 978-3-631-74971-5

MoE SR (1993) Metodické pokyny na vypracovanie dokumentov územného systému ekologickej stability. Bratislava. č. 101/93 – II

Ružička M, Miklós L (1982) Landscape ecological planning (LANDEP) in the process of territorial planning. Ekologia (CSSR) 1(3):297–312

Ružička M, Miklós L (1990) Basic premises and methods in landscape ecological planning and optimization. In: Zonneveld IS, Forman RTT (eds) Changing landscapes: an ecological perspective. Springer, New York, pp 233–260

Ružičková J (2002) Pripomienky k metodickým pokynom na vypracovanie projektov regionálnych a miestnych ÚSES. Interný materiál Aurex s.r.o., 3 pp

Šíbl J (1994) Metodika ÚSES a možnosti jej využitia pri tvorbe Európskej ekologickej siete EECONET v modelovom regióne Borská nížina. Kandidátska dizertačná práca. PriFUK, Katedra ekosozológie a fyziotaktiky, Bratislava, 156 pp

Chapter 4
Procedures of Designing the Territorial System of Ecological Stability

Abstract The chapter is the core part of the publication. The methodical procedure of TSES has several specific features as compared to other ecological networks. TSES in Slovakia focuses not only on the traditional frame of the ecological networks—as biocentres and biocorridors—but also on the proposal of whole-territory covering ecostabilisation measures, thus moving the ideas of ecological networks towards integrated management of optimum organisation and utilisation of the landscape as a whole. The methodical process consists five basic steps as analyses, syntheses, interpretations, evaluations and propositions. **Analyses** deal with all structures of landscape, namely with primary (abiotic), secondary (biotic–anthropic) and tertiary (socio-economic) structures, with specific regards to the real biota. The most important result of **syntheses** is the definition of potential representative geoecosystems for each model territory. The **interpretations** are oriented to definition of selective, localisation and realisation criteria for creation of ecological network, with specific emphasis on the interpretation of the ecological quality of the landscape structure. The **evaluations** are concerned with the definition of the suitability of geoecosystems to fulfil the functions of biocentres, biocorridors and interactive elements. The results of TSES are formulated in **propositions** which have several steps as:

- Delineation of frame of TSES: **biocentres, biocorridors** and **interactive elements**;
- Proposal of multifunctional **ecostabilising measures**, which should improve the spatial ecological stability and the environmental quality of the landscape as a whole. They are:
  - Location of new ecostabilising elements: groups and belts of multifunctional vegetation aimed to improvement of overall ecological spatial stability of the landscape, water retention, decrease of run-off, erosion and pollution,
  - Ecologically optimal land use of agricultural landscape: agrotechnical, agrochemical and agroameliorative proposals,
  - Measures for forest management: increase of diversity, revitalisation, considerate forestry techniques,
  - Measures within urbanised areas: improvement of the green infrastructure,
  - Elimination of **stress factors**, e.g. revitalisation of devastated areas, reduction of barrier effects.

L. Miklós et al., *Ecological Networks and Territorial Systems of Ecological Stability*, https://doi.org/10.1007/978-3-319-94018-2_4

- These “physical” proposals are completed by realisation measures, as proposals for legal protection of the TSES elements, passportisation of TSES elements and, as the final result, the formation of regulations for spatial planning procedures.

By combination of all groups of actions, the TSES becomes a whole-space covering tool for ecologisation of sectoral planning processes, as well as for integrated landscape management.

Keywords Analyses · Syntheses · Interpretations · Evaluations · Propositions
Ecostabilising measures

The presented publication does not aim to devise a brand new methodology of territorial systems of ecological stability, but to expound the existing principles of the methodological guidelines for the development of the regional and local TSES projects (Izakovičová et al. 2000) as real projects, although some chapters have been supplemented by the findings from previous research.

This methodology adheres to the original concept of TSES elaborated at the ILE SAS, legislative regulations, existing methodological guidelines for development of TSES, the experience resulting from implementation of specific projects as well as the above-mentioned guidelines drawn up by the Krajina 21 association in 2000 for the Ministry of Environment of the SR. The approach is based on a basic concept of interpreting spatial landscape-ecological stability as a dynamic ability of landscape structure to maintain spatial ecological relations within individual ecosystems for dynamic variability of conditions and life forms, even if that landscape is made up of local ecosystems with a different (and also low) level of ecological stability.

The methodological guidelines identify the following objectives for the compilation of TSES:

- Preserving and promoting development of the natural gene pool of the territory (biodiversity),
- Maintaining and designating stabilising elements in the landscape and sustaining their favourable influence on ecologically less stable landscape elements,
- Promoting a multifunctional approach to land use in order to protect the individual components of the environment,
- Preservation of important landscape elements,
- Elimination of stress factors, the factors hampering positive factors in the landscape.

These methodological guidelines partly draw on the previous methodological guidelines of the MoE SR from 1993. The following amendments have been introduced:

- Implementation of a new concept of potential representative geoecosystems for designation of new TSES elements,
- Interpretation providing more comprehensive description of the stability of landscape elements,

- Innovative assessment—the projection of a "problem map", a map of problems indicating the conflicts of positive and negative landscape elements,
- The proposals are more comprehensively formulated,
- Supplemented with passports serving as a basic information database,
- Supplemented with landscape-ecological regulations, which bridge the TSES documentation and other land planning documents.

4.1 Analyses

4.1.1 Transfer of the Results from a Higher to Lower Hierarchical Level of the Project

As mentioned in the theoretical–methodological section, the TSES is designed in a top-down manner. Therefore, the transfer of the results from a higher to lower hierarchical level is a prime and crucial step of the development of TSES as well as other different types of projects. It involves the application of the principles and skeleton of hierarchically superior projects, e.g. the General Plan of TSES. This phase also includes a description of location relations (geographical, geomorphological, bioclimatic, biogeographical and hydrogeographical position), basic spatial relations (natural barriers and corridors, natural gravitation, natural positional regions, ruggedness of the basic relief), tangible natural resources (geological, hydrological, soil, biotic resources), various intangible resources (hydropower), erosion, landslides, heat moisture potential, bioenergy potential, biodiversity and ecological stability of regions as a whole.

These natural changeable or partially changeable input indicators need to take into account also the partially changeable character of input socio-economic aspects, which influence the realisation of TSES; these aspects pertain geopolitical location of the region, socio-economic position, the basic land-use characteristics and the general state of the environment, economic development and the human potential of the region.

The analyses describe the major indicators of the landscape, which are essential for the development of TSES. The first stage of preparing the analyses includes compilation and subsequent assessment of relevant materials. Individual documents are compiled by means of excerpting the existing data relevant for a particular area and its landscape elements, interpretation of aerial images, processing of statistical data and field surveys. **Variability of the input materials depends on the hierarchical level of TSES**.

The following materials constitute the basic input data required for the creation of TSES:

- Geobotanical division of Slovakia—geobotanical map—provides the essential input for the projection of maps of representative ecosystems of the territory,

- Regional geomorphological division also represents an essential input for the preparation of maps of representative ecosystems of the territory,
- Climate zone map—database enabling analysis, synthesis and interpretation of abiotic landscape components,
- Geological maps contain the information on the substrate-soil conditions of the given territory,
- Phytogeographical and zoogeographical zoning—a key resource for basic biotic characteristics of the territory,
- Natura 2000 areas (habitats of European importance, protected bird areas),
- Representative geoecosystems,
- Databank of flora and fauna—a database of biologically important localities with a descriptions of species and their communities as groundwork data for the assessment of the areas of high gene pool importance,
- Forest stand maps—a basic database for the evaluation of forest ecosystems, and defining TSES elements on forest land,
- Databank of habitats—a database of important habitats as essential input data for the evaluation of real TSES elements,
- The results of measurements and monitoring of environmental components—as a fundamental basis for the evaluation of stress factors,
- Cartographic and textual materials on valuable landscape-ecological elements—documentation on nature protection, the General Plan of greenery development, documentation of individual departments (land-use planning documentation, forest management plans, water maps, hydroecological plans, bonited soil-ecological units)—as the basic input data to conduct a synthesis of the positive elements and abiotic complexes,
- Key physical–geographical cartographic materials—forming an essential part for the creation of abiotic complexes of the territory,
- Previously completed TSES documents (supra-regional, regional or local)—as the fundamental information database for identifying and delineating TSES elements,
- Aerial and satellite images—comprising the database for developing the current landscape structure (CLS) and identification of land degradation, waterlogging, etc.,
- Field survey—focused on updating the CLS elements, mapping of biota as well as selected socio-economic factors.

Several of the characteristics are projected in a map form in the Landscape Atlas of the SR (2002).

4.1.2 Analyses of the Primary Landscape Structure

This step includes the analysis of abiotic elements, namely: the analysis of relief (slope inclination, relief forms, ruggedness, types of relief), a substrate-soil and climate–hydrological conditions of the territory. An important part is the analysis of slope dynamics and drainage conditions. The analysis processes data that would facilitate designing of appropriately constructed spatial database—a synthetic map of abiotic complexes (Špinerová 2010). Assessing the conditions is important from the point of localisation of new TSES elements and proposal for ecostabilising measures.

(a) Relief

Relief is an intangible element of geosystem situated between the solid surfaces of Earth including the atmosphere and hydrosphere. From a mathematical perspective, it represents a continuous point field, and from the physical aspect it functions as an interface between solid, liquid and hard phase of geosystem. It is one of the key landscape differentiation factors. It largely shapes the horizontal and vertical flow of matter and energy. Therefore, the relief features dramatically affect the properties of other landscape elements. Their interpretation allows recognising the character and dynamics of the past and present processes occurring in the landscape. The aforementioned implies that detailed inspection of relief conditions is of particular importance.

There are a number of approaches which can be employed when processing the data of relief and their utilisation in the TSES documents. It can be claimed that, especially in case of small-scale documents, it is appropriate to operate morphometric parameters of the relief, whereas less detailed scales can also project morphographic and morphogenetic typological characteristics of the relief. Such characteristics can be obtained from the analysis of topographic maps on a scale of 1:10,000 and 1:50,000. In this case, it is also possible to adopt various standard methods, from manual to computer-assisted ones. Their detailed description is available in the following studies (Krcho 1973; Kudrnovká 1975; Miklós et al. 1997; Hrnčiarová and Miklós 1991; Minár 1992, 1995). Successful completion of this stage requires adequate professional experience.

(b) Geological bedrock and soil-forming substrate—rocks

Here, the term "rocks" refers to the topmost layer of the geological and soil-forming substrate. Rocks comprise of minerals, which build the solid layer of the Earth's crust. They differ based on their mineral composition, structure and texture. They are represented by hard and soft rocks, different types of weathered rocks and anthropogenic sediments. Rocks reflect long-term landscape evolution and influence other elements of the landscape as well as possibilities of its utilisation for not only technical but also bioproduction activities (as soil-forming substrate).

When processing data to be used in producing the MUSES and RUSES documents, it is necessary to identify lithological and genetic properties, chemical status and grain size composition of geological substrate units (Izakovičová et al. 2000). These properties can be obtained and interpreted from geological maps, engineering geological maps and other available documents. These include, for example, Atlas of engineering geological map of the SSR (1998), which contains maps at a scale of 1:200,000, basic geological maps projected at a scale of 1:50,000 and various thematic resources that are available also in the archives of the D. Štúr State Geological Institute.

(c) Soil

Soil constitutes a significant element of geosystem and fulfils indispensable environmental, ecological, bioproduction functions. It is a result of interplay of the atmosphere, hydrosphere, lithosphere and biosphere and therefore performs as an indicator of the present and partly also past landscape structure. The quality of soil cover is a factor determining the existence of certain types of flora and fauna in the landscape. It is also an irreplaceable natural resource with a production function, which is a critical factor determining the existence of human society.

When evaluating the data about soil for producing the TSES document, it is necessary to identify the typological characteristics of soils—soil type and subtype, soil state properties, especially grain size composition, depth and skeletality of soil. This information can be derived from the bonited soil-ecological units (BSEU) and other available materials. These documents, however, are not always accurate, and therefore they need to be reviewed and confronted with the data on other abiotic elements, such as relief and soil-forming substrate, which are closely intertwined with soil properties.

(d) Air and waters

Spatial and temporal variability of parameters of air and waters is a crucial factor in the differentiation of living systems, as well as life and human activities. Although the information on air and waters is monitored on a long-term basis, they usually have only a point character. The map projections based on these data are mostly of small scales.

When drawing up a TSES document, it is suitable to use either the typological characteristics of hydroclimatic data or selected state values, e.g. input of solar radiation, air temperature, the amount of atmospheric precipitation, prevailing wind direction and wind force, the occurrence of unfavourable conditions for dispersion of pollutants in the atmosphere (temperature inversions, windlessness). As for the state values of waters, depth of groundwater level, stream flow and drainage conditions can be utilised. Analytical data can be obtained from the Slovak Hydrometeorological Institute. They are also shown in the map form in the Landscape Atlas of the SR (2002). Data on the potential dynamics of solar insolation of relief can be interpreted from the morphometric parameters of relief (Krcho 1965, 1970; Miklós et al. 1997).

4.1.3 Analysis of the Current Landscape Structure/Land Cover Elements

As mentioned in the theoretical part of the publication, the secondary landscape structure is nowadays referred to as the current landscape structure (CLS) and forms **the core of the analysis for the creation of TSES**. The CLS represents a complex of physical landscape elements currently covering the Earth's surface (Ružičková and

Ružička 1973; Miklós and Izakovičová 1997). It is the "visible" layer of the landscape sphere, which shapes its general physiognomy. The elements of the secondary landscape structure (SLS) can be identified mostly as physical forms (ways) of land use and physiognomic–ecological formations of real biota (flora and fauna) and as man-made products (Miklós and Izakovičová 1997; Špinerová 2015). These types may perform several functions—land-use practices (Feranec and Oťaheľ 2001).

The analysis of the elements of the secondary landscape structure constitutes a basic framework for the identification and implementation of TSES elements.

CLS reflects the influence of human activities on the biotic and abiotic landscape elements, reveals the current state of biota, exposes the anthropogenic transformation of the landscape and informs about the use of the territory. The elements of the secondary landscape structure are physical landscape elements with a particular spatial delineation, so they are physically present elements covering the Earth's surface. For that reason, they can be seen as the elements of land cover and land use, habitats of animal species, urban–technical elements; the differences exist regarding their characterisation, the extent of their study and spatial projection. **The structure and detail of mapping units depend on the level of TSES, character of the studied territory and the scale of projection**.

In the following paragraphs, we will describe the SLS from

- The physiognomic–functional perspective—as simplified mapping unit of CLS,
- The biological–ecological perspective.

4.1.3.1 Mapping the Units of the Current Landscape Structure/Land Cover

CLS units are mapped based on their physiognomic–functional characteristics in the following groups (in accordance with the classification by Ružičková and Ružička 1973):

- Forests

The group comprises all types of forests, from large forest complexes to small-scale forests, indigenous or altered. They can be characterised by various indicators, e.g. physiognomic–ecological indicators, composition of tree species, dominating species and their combinations, origin, vertical structure, their functional integration in the landscape, age, phytocoenological or forest-typological indicators. When creating the TSES documentation, the data on forest vegetation draws on the information about groups of forest types and stand types contained in forest stand maps and forest management plans.

- Non-forest woody vegetation (NFWV)

This group includes scattered non-forest woody vegetation in the landscape, in particular, accompanying vegetation of roads, watercourses, field baulks, boundary lines, groves, solitary trees, shelterbelts, shrubs and their clusters. The basic characteristics involve the following indicators: density, continuity, length, covering area. The listed indicators can be interpreted from up-to-date satellite images or topographic maps; the species composition of NFWV is identified by field survey.

- Permanent grasslands

The group contains pastures, meadows, as well as other natural and semi-natural herbaceous communities. The basic assessment criteria of CLS comprise the used permanent grasslands and the degree of anthropic impact. From this aspect, they can be classified into two main groups—extensively and intensively used pastures and meadows. Specific mapping units are detailed by means of in-depth geobotanical investigation.

- Arable land and permanent crops

The group includes large-block and small-block arable land with sown temporary sawn grassland and crops. In some territories, they represent the CLS elements which cover the largest areas. This category also encompasses vineyards, orchards, hop fields and gardens. The intensity and the ways of their use are considered the basic characteristics.

- Mosaic structures

Under this complex, we map groups of trees, shrubs (usually not occurring on forest land), meadows, pastures and small-block arable land. They are spread over large areas and comprise various proportions of woody species.

- Water areas and watercourses

The category includes all kinds of artificial or natural water areas and lines. They are characterised by the degree of indigenousness, functional use, etc.

- Elements without permanent vegetation cover or only its initial stages

They are represented by natural (rocks, sand, gravel deposits, etc.) or man-made, currently often naturalised elements (e.g. abandoned quarries and clay pits) and devastated areas and the like, with minimum soil cover and corresponding vegetation. Their defining characteristic is their shape and structure. Based on this, they can be categorised into gorges, cliffs, cirques, slope debris, riverside gravel and sandbars, sand dunes, dumps, ravines, quarries, sandpits, etc.

- Anthropogenic technical elements

These elements comprise a relatively extensive group of various human-made elements. They can be characterised based on their functional use and include:

– Industrial production areas—various industrial structures and their adjacent warehouses. They can be subdivided by the type of the industrial production (engineering, metallurgic, food, chemical, consumer goods industry and so on.), size and shape of a structure, the intensity of their negative impact on the landscape, etc.,

- Mining areas—anthropogenic structures in the landscape designed for the purposes of extraction of mineral resources. Their location within the landscape is determined by the occurrence of the mineral resource. Based on the character of mining, they can be further divided into groups—surface mines, underground mines, mining pits, etc.,
- Pipelines—anthropogenic linear elements designed to transport materials, energy and information. Based on their localisation, they can be categorised into underground (pipelines, cable lines, water lines, etc.), aboveground running just above the Earth's surface, e.g. gas line) and aerial (running in the air—e.g. power transmission lines, telephone wires etc.),
- Transportation elements—these elements provide transport of persons, energy, materials and information. They can be categorised in terms of the character of transport (road, rail, air, water), society-wide importance (corridors of world importance, European importance, national importance, etc.). From the physiognomical point of view, they can be divided into point elements (petrol stations, etc.), lines (transport corridors, road, railway and water lines) and areas (airports, ports, etc.),
- Agricultural areas—structures intended for agricultural production (livestock production buildings, storage areas, mechanisation facilities, fertiliser storages, field dung heaps, etc.). They are evaluated based on their use and size, or the degree of their negative impact,
- Forest management technical elements—technical elements related with forest management activities (timber yard, mechanisation centres, sawmills, etc.),
- Water management areas—designated for the protection and use of water resources. All water management structures and buildings—amelioration elements, pumping stations, water treatment plants, etc., are subjects of mapping,
- Residential and recreational areas—residential areas, areas of services, holiday cottages, cottage settlements, camps, residential areas, areas of services, linear recreational elements such as hiking trails, nature trails are projected on a map,
- Other technical elements—e.g. military areas (the CLS elements performing the activities related to state defense and can be classified on their function, e.g. military training site, barracks), waste dumps (waste substances from households—the main assessment criterion is the type of waste), building sites and others.

CLS constitutes a basis for designation of current significant landscape ecostabilising segments and determines spatial distribution of stress factors, character of barriers restricting and threatening the ecological stability and quality of a territory.

The essential part of the CLS is also the inventory and identification of the NFWV structures as well as the assessment of current ecologically significant landscape segments with respect to the characteristics of present flora and fauna. An example for the CLS key for MUSES, the lowest, most detailed level is given in Table 4.1.

Table 4.1 Example of the key for mapping units of the current landscape structure for the local TSES

	Groups of CLS elements	CLS elements
(A) *Forest vegetation*		
1.		Oak forests
2.		Hornbeam–oak forests
3.		Pine–oak forests
4.		Birch–oak forests
5.		Birch–alder forests
6.		Hardwood floodplain forests—hornbeam–ash floodplain forests
7.		Transitional floodplain forests—oaks–ash floodplain forests
8.		Transitional floodplain forests—elm–ash floodplain forests
9.		Softwood floodplain forests—willow–poplar forests
10.		Birch–oak forests
11.		Oak–beech forests
12.		Forest-steppe–pine forests
13.		Oak–pine forests
14.		Spruce–pine forests
15.		Ash–alder forests
16.		Beech forests
17.		Fir–oak forests
18.		Oak–fir forests
19.		Fir–oak forests
20.		Spruce–pine forests
21.		Oak–pine forests
22.		Beech forests with admixed fir
23.		Beech forests with admixed oak
24.		Oak–fir forests
25.		Mountain beech forests
26.		Spruce–fir forests
27.		Maple–fir forests
28.		Beech trees with admixed fir
29.		Fir–maple forests
30.		Pine–spruce forests
31.		Fir–beech forests
32.		Fir–beech forests with admixed spruce
33.		Mountain spruce forests
34.		Alpine spruce forests

(continued)

Table 4.1 (continued)

	Groups of CLS elements	CLS elements
35.		Dwarf pine
(B) *Non-forest woody vegetation (NFWV)*		
36.		Groups of natural NFWV
37.		Groups of planted NFWV
38.		Riparian stands
39.		Line natural NFWV
40.		Line planted NFWV
(C) *Permanent grasslands*		
41.		Vegetation of stagnant water, bogs and abandoned channels
42.		Vegetation of moors and mires
43.		Wet lowland meadows
44.		Wet foothill and mountain meadows
45.		Extensively used meadows and pastures
46.		Intensively used meadows
47.		Intensively used pastures
48.		Dry meadows and pastures
49.		Subalpine vegetation
50.		Alpine vegetation
(D) *Arable land*		
51.		Large-block arable land (fields)
52.		Large-block arable land with low occurrence of NFWV
53.		Large-block arable land with occurrence of NFWV
54.		Small-block arable land—narrow-stripped fields
55.		Small-block arable land—terraced fields
(E) *Permanent crops*		
56.		Large-block vineyards
57.		Terraced vineyards
58.		Small-block vineyards
59.		Abandoned vineyards
60.		Orchards and plantations
61.		Abandoned orchards
62.		Hop fields
63.		Gardens
64.		Nurseries for ornamental and fruit trees
65.		Uncultivated agricultural areas (fallow lands)

(continued)

Table 4.1 (continued)

	Groups of CLS elements	CLS elements
(F) *Mosaic structures*[a]		
66.		Mosaic of small-block arable land, meadows and NFWV
67.		Mosaic of meadows and NFWV
(G) *Watercourses and water areas*		
68.		Watercourses natural (unregulated)
69.		Watercourses regulated
70.		Oxbow lakes and meanders
71.		Channels
72.		Natural water surfaces (lakes, mountain lakes)
73.		Dams
74.		Abandoned gravel pits
75.		Ponds
(H) *Elements without permanent vegetation cover, or at its initial stages*		
76.		Natural rock formations
77.		Natural sand and gravel deposits
78.		Quarries
79.		Gravel pits, sandpits, clay pits
80.		Overgrowing quarries
81.		Overgrowing sandpits, clay pits
82.		Other areas without permanent vegetation cover, or at its initial stages
(I) *Industrial and mining areas*		
83.		Factory premises (shop floors, warehouses, a.o.)
84.		Structures of surface mining
85.		Structures of subsurface and underground
86.		Industrial waste dumps
87.		Waste ponds
(J) Pipelines		
88.		Extra high-voltage power lines
89.		High-voltage power lines
90.		Medium- and low-voltage power lines
91.		Pipelines[b]
(K) *Traffic elements*		
92.		Highways
93.		I. class roads
94.		II. and III. class roads

(continued)

Table 4.1 (continued)

	Groups of CLS elements	CLS elements
95.		Unpaved roads
96.		Railways
97.		Cog railway
98.		Railway areas
99.		Parking areas
100.		Airports
(L) *Agricultural elements*		
101.		Agricultural production areas
102.		Field airports
103.		Field dunghills
(M) Residential areas		
104.		Residential areas with low-rise buildings
105.		Residential areas with high-rise buildings
106.		House gardens
107.		Civic amenities
108.		Parks
109.		Cemeteries
110.		Recreational and sports facilities and areas
111.		Cottage and log cabin villages
112.		Caravan parks and campgrounds
113.		Spa areas
114.		Hiking trails
(N) *Other areas*		
115.		Military facilities and areas
116.		Waste dumps
117.		Building sites
118.		Non-used land

Note [a]mosaics can be more accurately characterised based on the distribution of NFWV; [b]pipelines can be further subdivided according to their use

4.1.3.2 Biotic–Ecological Characteristics of the Current Landscape Structure/Land Cover

Biotic analyses of TSES may lead to further valuable syntheses, special-purpose interpretations, comprehensive classification and contributive proposals, and therefore **they need to be underpinned, at least at the local level, by detailed integral field surveys**. That is the only way to appreciate the real state of the biota within the concerned territory.

Analyses of biotic conditions in the project of territorial systems of ecological stability comprise of phytogeographical and zoogeographical division of the territory, characteristics of potential natural vegetation and real vegetation, habitat types (delineation and description), migration routes of animals, habitats of rare, protected, endangered and invasive species of biota.

Full area field surveys help to identify and delineate the habitats which are described by the following aspects: physiognomy, species composition, ecotopes and their occurrence in the studied territory, especially in terms of well-established habitat catalogues (Ružičková et al. 1996; Stanová and Valachovič 2002).

The biotic–ecological description of CLS elements for the local TSES aims to map and assess plant and animal species and their communities occurring in the given territory. The analyses should provide quantitative and qualitative characteristics of flora and fauna of forest communities, non-forest woody vegetation, permanent grasslands (description of meadow communities), wetlands and other habitats, gene pool areas and other major natural elements in the territory.

(a) ***Quantitative description*** addresses the spatial representation and distribution of landscape elements, their area and spatial parameters (surface area, shape, width and length, location, distance from the closest similar element, gap size between line formations, height of a vegetation layer, etc.). Spatial differentiation of biotic elements is a part of CLS evaluation.
(b) ***Qualitative description*** draws on a CLS map supplemented by the results of a relevant field survey with regard to required detail and scope of its projection. The procedure presupposes conducting an intensive field survey with the purpose of updating already existing data on current vegetation cover and fauna occurring in significant habitats, (the data can be obtained from existing databases, documentations, from specialists, in the field) and subsequent map projection of distribution of systematic groups their communities with additional qualitative data.

The qualitative description of vegetation formations examined during field survey focuses particularly on the species distribution in each vegetation layer (E3, E2, E1), coverage of plant species or layers and their total coverage, canopy of tree and shrub layer, connectivity (interconnection with adjacent vegetation formations, their length and a number of gaps they contain, type of barriers, etc.), representation of communities and types of habitats and their current state (the most widespread, naturally rare—unique, threatened and representative within the concerned geoecosystem), age structure (non-forest woody vegetation, forest stands), genesis of vegetation forms and health condition of trees (Diviaková 2010).

Descriptions and evaluations of forest communities can draw on the stand and typological maps, but also other parts of the forest management plans, in particular:

- Stand description and plan of management actions,
- Results of the assessment of the ecological stability of forest stands based on the analysis of naturalness and origin of wood species composition and age of stands, etc.

It is crucial to conduct a thorough field survey to adequately perform the assessment of habitats of non-forest woody vegetation (e.g. riparian stands, shrub habitats) and permanent grasslands (meadow and pasture habitats), as the current data for the territory of Slovakia is insufficient (except maybe habitats of European significance).

Additional information concerning this part can be gained, apart from the above-mentioned forest management plans, also from text and map materials such as management programmes for protected areas, reservation books of protected areas. Valuable information on flora and fauna is provided by relevant databases (Pauditšová et al. 2007).

As for animal species, within the framework of the local TSES, it is important to pay close attention to the evaluation of invertebrates, amphibians and reptiles, birds, fish and mammals and gather the following information on:

- Regionally significant populations of endangered and protected taxons,
- Species protected under international conventions and the European Union directives on habitats,
- Taxons, whose viable populations require for their existence a large territory and whose protection cannot be ensured only by measures implemented at the local level,
- Nesting and wintering populations of endangered and protected taxons and data on migration routes of regional and higher importance, etc., at the regional level and local migration routes at the local level,
- Migration routes and barriers in the evaluation of mammals,
- Migration routes, spawning sites of fish, hatching sites and wintering sites,
- Amphibians and reptiles at the local level, barriers and road sections with higher occurrence of collisions with vehicles, potentially suitable locations for restoration (revitalising) of significant habitat or for establishing substitute habitats.

For the purpose of the RTSES and LTSES, it is advisable to evaluate especially vertebrates and selected groups of invertebrates with a relatively higher bioindication value and properly researched in Slovakia and central Europe, because it permits the comparison of importance of particular regions, biocentres and biocorridors for particular animal species (groups). The assessment of biota should map selected types of habitats (Table 4.2.).

The data acquired should form a basis for a database of all **ecologically significant landscape segments**, not only those currently protected by law. Identification of ecologically significant landscape elements forms the core of the TSES skeleton.

4.1.4 Analysis of the Tertiary Landscape Structure—the Socio-economic Factors

The socio-economic factors (SEF) enjoy a special status within the framework of the TSES methodology, because they encourage as well as threaten the ecological stability of the territory. Analyses of the socio-economic factors can be divided into two major groups (Hrnčiarová and Izakovičová 2000):

Table 4.2 Example of mapping groups of animal species for selected habitat types

Habitat type	Mapping group
Permanent grasslands	Saltatoria—grasshoppers, locusts,
Stagnant waters	Dragonflies (Odonata), Crustaceans (Crustacea)
Running waters—benthic organisms	Crustaceans (Crustacea), Stoneflies (Plecoptera), Mayflies (Ephemeroptera), Caddisflies (Trichoptera)
Forest habitats	Capercaillie, woodpeckers, cavity-nesting birds, wolf, lynx, bats (wintering and summer colonies in caves and forests and hunting grounds)
Wetlands and water habitats	Waterbirds, beaver, otter
All types of habitats	Molluscs (Mollusca), spiders (Araneidea), beetles (Coleoptera), butterflies (Lepidoptera), birds of prey and owls (nesting areas and hunting grounds)

4.1.4.1 Socio-Economic Factors Arising from the Protection of Nature, Natural Resources and Cultural Heritage—"Landscape Priorities"

These socio-economic factors promote the overall functionality of TSES; hence, they are defined as a "landscape priorities", "positive elements". They include the socio-economic factors (SEF) resulting from the protection of natural resources and cultural heritage and enhance their legal status, but also map ecologically important landscape elements without legal protection. The outcome of this part is a specification and method of spatial projection of protected areas and their buffer zones at various levels, as well as the account of other currently unprotected significant habitats. They form a group of elements with a positive or neutral impact on the landscape. Their definition, especially the legislative one, is not considered an absolute prerequisite for permanent sustainable ecological stability (e.g. degradation due to exposure to external factors), but rather as an indicator of ecological and cultural values in the landscape (Hrnčiarová 1998; 1999a, b). The elements are represented by an array of unique natural, cultural and social values that improve the mental, health and cultural development of man and his living space, but also other biota in the landscape. They also ensure the preservation of natural resources, gene pool, ecological stability, landscape diversity, etc. These elements exist as various plant and animal formations—natural factors, whose significance is given by various levels of legal protection. They cover only a specifically selected part of the Earth's surface (nature protected areas, elements of supra-regional and regional TSES, protective forests, special-purpose forests, protected water areas, zones of sanitary protection of water resources and other), and therefore it is necessary to ensure their proper functionality in the future. They can be subdivided as follows:

- *SEF of nature conservation*: categories of territories with unique values and landscape elements, which are protected by law because they possess some distinctive and extraordinary qualities. They require special care and protection, and their significance for man lies in numerous useful functions they perform, e.g. ecostabilising, gene pool, anti-pollution, anti-emission, hygienic, historical, aesthetic a.o. Their properties may change over time, and therefore even their legislative framework is not fixed and unalterable. There are important factors restricting the development of activities in the landscape. All the above-mentioned elements should be incorporated in the TSES documentation as TSES elements in the form of protected nature areas (territorial protection), protected species (species protection), protection of woody species (protected tree), protection of the territory under international treaties, territorial systems of ecological stability (existing TSES materials).

- *SEF of natural resources protection*—categories of territories with the occurrence of natural resources which carry out important socio-economic—utility as well as ecological functions in the landscape. There are legislative provisions in place to ensure the conditions for the maintenance, renewal and rational use of natural resources, but also the preservation of gene pool, biodiversity and ecological stability in the territory. Some of these territories can be immediately incorporated as TSES elements; some contribute towards creating favourable ecostabilising conditions. At the same time, they constrain a number of development activities in the landscape. They include: the protection of forest resources, protection of water resources, protection of natural healing resources, protection of natural spas and climatic conditions with salutary benefits, protection of land resources, protection gene pool resources, protection of mineral resources.
- *SEF of the protection of monuments and historic sites*: TSES focuses mainly on a special group of cultural heritage structures (movable and immovable monuments) that occupy the landscape-ecological relevant areas and are recorded in the central list of cultural monuments. Any set of immovable cultural heritage monuments can be declared a historic reserve. Protective zones need to be delineated for the protection of such monuments. A historic zone is a special category declared as a territory with a historical residential arrangement, or its part, with lower proportion of cultural heritage monuments or historic environment, or a part of landscape unit exhibiting cultural values. A protected area can be delimited based on archaeological finds, which means any territory that provides evidence of human life and activities.
- The positive landscape elements form the basis for delineation and proposal for TSES elements. A different projection scale for the regional TSES and local TSES does not allow mapping the same units. The local TSES (LTSES) permits mapping of all presented elements, whereas the regional TSES (RTSES) elements must be subject to selection process.

4.1.4.2 Socio-economic Factors Threatening the Functioning of Territorial System of Ecological Stability—Stress Factors in the Landscape

These factors are defined as "stress factors, negative elements". They include the socio-economic factors resulting from human activities and plans in the landscape, which adversely affect natural development of ecosystems, but also create potential landscape barriers which disrupt the spatial continuity of TSES. They can be represented by tangible and intangible products of human activity. Based on their genesis, they can be divided into two groups (Izakovičová 1996a, b, 1997, 2000):

- Primary stress factors—primary causes of stress. The group comprises semi-natural and artificial physical anthropogenic elements (industrial and agricultural buildings, road areas and lines, surfaces and structures of agriculture and forestry, residential and recreational areas, etc.). These stressors are characterised by their distinct spatial manifestation in the landscape. The mere presence of a stress factor in the landscape already causes stress which, from the point of TSES, then leads to changes in landscape structure and its use (loss of natural ecosystems due to the pressure of anthropic activities) and impedes the flow of genetic information, migration of biota. The material substance of the primary stress factors is mapped in the form of CLS elements.
- Secondary stress factors—negative factors accompanying human activities in the landscape which lack a clear-cut spatial delineation, and therefore their spatial projection varies. The negative effect of these factors is manifested as threats, disturbances or distortions of the conditions of ecosystem development. They can be collectively termed as deterioration (degradation) stressors. The territories of their activity are demarcated as buffer zones of technical structures and labelled as sanitary protection zones of technical elements or security buffer zones. The second group of these factors denoted as deterioration factors is delineated directly in the territory of their occurrence.

TSES projects classify the socio-economic factors (SEF) which are bound to secondary stress factors into two main categories:

(1) *Sanitary protection zones, buffer and safety zones*

Sanitary protection zone (SPZ) is usually demarcated around technical elements in order to protect the surrounding area from their adverse effects. They can be regarded as zone of the negative impact of the technical structure on the environment. They include (Izakovičová et al. 1997):

➢ *SPZ of industrial and storage facilities*—areas designated in the vicinity of factories, whose negative impact determines the size of such areas. Their negative effect is observable in shrinking the areas of natural ecosystems, but also production of emissions, solid, liquid and gaseous waste, increased noise and dust levels, etc. The nature of the negative impact is determined by the type of industrial production. These buffer zones dramatically constrain development of other production or non-production industries, as they restrict performance of certain activities within their territories. They have also negative impact on the TSES elements located in their close proximity. Sanitary protection zones must be established around factories and industrial facilities spanning the following widths:

- Class I. zone (over 500 m)—severely disruptive or threatening production processes,
- Class I. zone (100–500 m)—medium disruptive or threatening production processes,
- Class I. zone (up to 100 m)—moderately disruptive or threatening production processes.

➢ *SPZs of landfills* are situated around landfills extending 300–500 m. Their purpose is to protect the immediate environment exposed to the negative impacts of waste dumping such as dustiness, bacteriological contamination, sources of emissions, stench. As was the case with the previous SPZ, these areas restrict performing of certain socio-economic activities. From the health, hygiene and landscape point of view, it is advisable to plant an insulating vegetation strip around a landfill, which would mitigate its negative environmental impacts.

➢ *SPZs of wastewater treatment plant (WWTP)* range from 100 to 1000 m and are delineated around wastewater treatment plants on individual basis, depending on the population and the degree of treatment. The goal of the zone is to protect the surrounding environment of the treatment plant from discharge of harmful substances generated during the treatment process.

➢ *SPZs of agricultural areas* are established in the vicinity of these structures spanning from 300 to 1000 m to provide protection from negative influences such as noise, dust, stench. The main demarcation criteria include the type and size of the livestock population, as well as the method of collection, disposal and use of excremental materials.

➢ *Protection zones of military structures and military protection zones*—these zones restrict performing all agricultural activities, because the defensive-strategic function of the structures determines their strict monofunctional use. Their negative impacts on the environment are manifested in seizing natural ecosystem space, disrupting and threatening vegetation cover as well as increased noise and dust levels.

➢ *Protection zones of linear technical elements* are designated to protect these elements and their surroundings with a special management plan which restricts performing of certain activities. This relates to the following protective zones:

- *Railway protection zones*—formed by 30–60 m land strips on both sides of railway. Any landscape management activities must be in line with the railway protection and cannot threaten or restrict the normal operation of the structure. The negative effects result from the development of rail transport; they are in particular noise, production of emissions, dust, etc.

- *Road protection zone* is established to protect roads and traffic. Any use of the plots situated within such zones must be in accordance with this goal. The boundaries of road protection zones are demarcated by vertical surfaces located 15–100 m on both sides of a road. These zones represent the areas of negative impact of road transport on the environment. The major negative impacts caused by the development of road transport are noise, dust, production of emissions, barrier effect for migration of biota, light effects, etc.
- *Protection zones of power lines* are established as strips along the lines with a width of 25 m for extra high-voltage lines, 20 m for high-voltage lines and 15 m for low-voltage lines. In these zones, it is restricted to erect buildings or perform any surface treatment activities that could disturb the territorial stability, build facilities or plant trees, which could potentially endanger proper functioning of power facilities. From the ecological point of view, these elements act as barriers for migrating bird populations. Another noteworthy influence is their electromagnetic effects on living organisms.
- *Protection zones of gas facilities* are designated along both sides of a pipeline, and their width ranges from 10 to 50 m. In these zones, it is necessary to exclude activities with the potential to endanger the operation of the facility (groundwork, refuse dumps, storage of inflammable substances, etc.). From the landscape-ecological point of view, the presented lines pose barriers for development of soil biota.
- *Protection zones of cable lines* aim to protect cables and their associated facilities. The protection zone of cable lines is 2–3 m in width. All activities which could threaten cables and compromise their safe operation (such as deep ploughing, excavations and other groundwork, drainage) are prohibited in this protection zone. Moreover, it is forbidden to construct buildings, warehouses, landfills, etc., as their negative effect is similar as in case of the previous group.

The assessment of the protective linear technical elements is essential from the localisation of the TSES elements, because these zones function as spatial (implementation) constraints of the distribution of TSES elements.

(2) *Deterioration factors*

As indicated before, these negative factors accompanying human activities in the landscape often lack accurate spatial delineation and their boundaries may shift. The following are the key groups of secondary stress factors (Izakovičová 1995):

- *Contamination of rock environment* represents natural and anthropic accumulation of components exceeding the admissible amounts for the given lithotype,
- *Air pollution by heterogeneous substances*—areas in which pollutants exceed the limit (emission, immission and deposit limits),
- *Noise load of the environment*—environments in which the noise level exceeds acceptable levels,
- *Odour load of the environment*—a phenomenon which can be objectively measured only with difficulty and is mostly subjective,

- *Vegetation damage*—areas with observable changes caused by natural, semi-natural or anthropogenic factors, especially by emissions and are assessed according to the degree of damage,
- *Surface and underground water pollution*—water with excessive content of heterogeneous substances—water with excessive content of contaminants,
- *Soil degradation*—areas with disturbed physical, chemical and biological properties of soil.

Mapping and assessment of stress factors are a necessary part of TSES because stress factors act as:

- Physical barriers for the elements of the territorial system of ecological stability,
- Initiators of environmental problems,
- Environmental regulations of (limits and restrictions) spatial development.

4.2 Syntheses

The main aim of syntheses is development, description and classification of homogeneous spatial units with clearly defined indicator values of the properties of elements processed in the synthesis. Thus, the resulting spatial units hold the same indicator values of properties at the given differentiation level across the whole territory of the unit. These final units are abiotic, biotic or socio-economic or landscape-ecological complexes of which have various combinations of property values of individual elements. The synthesis is commonly conducted by superposition of analytical or interpreted maps. Created complexes are then categorised by their properties combination into typological units, creating types of complexes, for example, types of abiocomplexes—ABK types.

In the framework of the TSES, the synthesis procedure comprises of four basic groups:

- Identification and overview of potential representative geoecosystems,
- Synthesis of abiotic complexes,
- Synthesis of socio-economic factors that support functioning of TSES,
- Synthesis of socio-economic factors that threaten functioning of TSES.

4.2.1 Potential Representative Geoecosystems

4.2.1.1 Methodology and Criteria for Establishing Potential Representative Geoecosystems

The methodology of delineation and description of potential representative geoecosystems (REPGES) is adequately explained in the following publications Miklós et al. (1997) and Izakovičová et al. (2011), so here we will only deal with the fundamental criteria of REPGES characteristics.

As mentioned in Sect. 2.5.2, identification and definition of potential representative geoecosystems have **strategic importance** for nature conservation, as well as for the creation of TSES—it serves as a **conceptual scheme** for a systemically formulated plan to determine:

- Which representative conditions and forms of geoecosystems occur in a given territory—and what needs to be protected and preserved,
- Which are already preserved and protected, and the resulting findings,
- Which are preserved, but unprotected, as well as,
- Which ones are and which are not physically preserved in the given territory.

In practical terms, these outputs should serve as an ecologically founded basis for **proposals for new protected areas**. However, such strategic analysis and delimitation of REPGES are not conducted for each TSES, and biocentres are established not according to the representative conditions of the concerned territory, but simply considering just the current state of CLS, so biocentres are ultimately represented by CLS elements with a similar current state. This can lead to many geoecosystems not being protected by law and being excluded from the TSES.

Selected criteria for identification and characteristics of REPGES are as follows:

(a) *Types of geoecosystems—content criterion*

Standardly, two models are applied in this case:

- Model of natural heritage (natural-cultural-historical approach). This is a model of preserving real geoecosystems.
- Landscape-ecological model

The real landscape necessarily contains geoecosystems which are vital for human life and whose vegetation cover is partially or entirely anthropogenically altered (e.g. agroecosystems, urban systems), but from the point of the conditions for life—-geoecodiversity—they are still considered important and functional. Therefore, the main criterion of this approach is functionality, e.g. ecological functions (functions of reproduction, shelter, food, migration) in a particular landscape, bioproduction, soil and water protection, hygienic, climatic functions, impact on adjacent geoecosystems (ecostabilising functions) and other so-called socio-ecological functions (Jurko 1990). Thus, the question is how a given geoecosystem functions rather than what kind of geoecosystem it actually is. Therefore, in the second approach worthy-of-preservation and preservation-requiring geoecosystems represent also those whit modified biota. This is the model of large-area preservation of **potential geoecosystems** (what "type" of geoecosystem).

Note: Whereas the model of preserving real geoecosystems is traditionally the approach of typical nature protection (albeit does not apply all cultural-historical aspects to all geoecosystems), the model of preserving potential geoecosystems is an approach of modern whole-territory nature and landscape protection with dominating landscape-ecological approach. Furthermore, it is noteworthy that preservation of a certain type of potential geoecosystem allows development of a large number of real geoecosystems.

(b) *Surface area*

The surface area of geoecosystems is a critical criterion of representativeness because the geoecosystems which cover larger areas are considered **positively** representative. This does not mean that small-area geoecosystems cannot be designated as representative, but large-area, "commonplace", anthropogenically affected geoecosystems oftentimes fail to attract enough interest to be protected, and this may, in the long run, cause their permanent degradation and subsequent ignoring of their contribution to general biodiversity. It should be emphasised that the extensive geoecosystems play a decisive role in maintaining all ecological relations, from bioproduction to environmental and ecological functions. Small-area geoecosystems can be also considered representative if:

- An incidence of more small-scale geoecosystems creates a specific landscape mosaic, which ultimately becomes extensive (e.g. alternation of thermophilic oak-woods and oak–hornbeam woods in basin hill lands).
- The occurrence of small-area geoecosystems causes large geoecosystems to acquire a specific spatial structure and characteristic features (furrows in highlands, canyons in plains).
- From the geological genesis point of view, they cover a priori small area (e.g. Klippen Belt in the Western Carpathians), but are specific for a particular region.

Other small geoecosystems—equally important for maintaining geoecodiversity—may be considered representative, especially at the lower hierarchical levels, in detailed studies or components of larger representative geoecosystems.

(c) *Hierarchical level of representative geoecosystems*

Determination of hierarchical levels is the **first step** in the identification of representative ecosystems. Form the systemic perspective, geoecosystems have all topical, choric (regional) and planetary dimension, and thus the criteria for the creation of TSES should be assessed within a particular hierarchical level.

Nowadays, territorial systems of ecological stability recognise supra-regional, regional and local level, which is in accordance with the three levels applied in spatial planning, which includes the TSES as its integral part.

Based on the aforementioned criteria, the REPGES concept in Slovakia can be described as follows.

4.2.1.2 Representative Geoecosystems in Slovakia

The **regional** level is the most characteristic level of the representative geoecosystems in Slovakia. However, to fully appreciate reciprocal relationships, it is necessary to take into account also supra-regional and local level of REPGES.

From spatial point of view, REPGES can be characterised from the following perspectives:

- From a typological perspective approximately by the units which are at the level of mapping units of the geobotanical map (Michalko et al. 1986) and at the level of the types of landscape-ecological complexes of the SR (Atlas krajiny SR 2002),
- From the regional perspective at the level of phytogeographical subdistricts or subunits of the geomorphological division of Slovakia (Atlas krajiny SR 2002).

The strategic objective of the process of defining a representative ecosystem is to:

- Determine a REPGES type of each region,
- For each REPGES type, determine a region in which it is representative, in accordance with the rule that every region should have a REPGES and that each REPGES type should be somewhere representative. This constitutes a precondition of full-scale preservation of diversity and life forms—geodiversity.

(a) *Types of REPGES*

REPGES types are defined by:

- **Zonal** (bioclimatic) conditions expressed by vegetation zones. It should be noted that it is zonality caused azonally—by altitude.
- **Azonal conditions**—primarily by geological base and relief, secondarily by waters, soils as well as secondary zonality—mesoclimate affected by relief.

(a1) **Zonal conditions of REPGES types**

These are bioclimatic conditions which are apparent in vegetation cover and zonality of soils. This manifestation can be assessed directly in case of near-natural vegetation and indirectly by the character of soils in case of altered vegetation or agroecosystems. In many instances, however, we infer bioclimatic conditions and potential natural vegetation solely based on the land surface, not vegetation cover. Therefore, in such instances it is only an interpretation of abiotic conditions for the purpose of typification of living systems.

Bioclimatic conditions are comprehensively presented in the geobotanical map of the SR (Michalko et al. 1986) and other cited studies mentioned above.

(a2) **Azonal conditions of REPGES types**

These conditions are assessed based on the relief and geological base (drawing on available materials of relevant research projects and the maps from the Atlas of the SSR) as they reflect the most defining features of hydrologic conditions and soils, and azonal communities of potential natural vegetation. These conditions are further described in the above-mentioned specialised studies.

(a3) **Complex characteristics of REPGES types of the SR**

Zonal and azonal conditions are observable in real REPGES types simultaneously, so none of these aspects is superior. The aforementioned classification is necessary for better orientation in REPGES types and facilitates their organisation. A comprehensive list of REPGES types is presented in the Landscape Atlas of SR (2002) and in the Atlas of representative ecosystems of Slovakia (Miklós et al. 2006), representative landscape types of Slovakia (Bezák et al. 2010) and model of representative geoecosystems at the regional level (Izakovičová et al. 2011).

REPGES types have a character of potential geoecosystems because they are delineated on the basis of the **abiotic** conditions that demonstrate certain potential for the development of geoecosystems and are also characterised by means of potential vegetation.

In a real landscape, most of REPGES types can be identified by several variants of current situation:

- Forested by quasi-natural forest or planted monocultures,
- Non-forested, with quasi-natural grasslands, permanent crops (vineyards, gardens, orchards, parks), with agricultural crops on arable land,
- Built-up areas.

Geoecosystems that are far from the state determined by potential vegetation are also important for preservation of biodiversity. Therefore, the strategic objective of REPGES identification should focus on preserving as many variants of REPGES types as possible. On the other hand, it should be stressed that at a certain hierarchical level and scale of REPGES map projection, it is unsuitable to demarcate some REPGES variants.

(b) *Regions of REPGES*

The regional aspects of geoecosystems are expressed in biogeographical division, usually in phytogeographical units, which constitute the basis for the regional division of REPGES (map of phytogeographical division, the Landscape Atlas of the SR 2002; Hrnčiarová et al. 1991).

Regions of REPGES are delineated by a spatial synthesis of the phytogeographical units and the units of geomorphological division of Slovakia. These are equal to a hierarchical degree of "subunit" in the geomorphological division of the SR. The regional geomorphological division of Slovakia is quite detailed, so also **subregions** and their **parts** can be effectively used for the purposes of TSES. *It needs to be noted that in case of biogeographical division, the boundaries of the units cannot be seen as precisely fixed. It is then tolerable to adjust the phytogeographical boundaries of subdistricts to the boundaries of regional units of the geomorphological division, whose boundaries are much more precisely drawn.*

The REPGES regions are presented in table and map form in the Landscape Atlas of SR (2002) and in the Atlas of representative ecosystems of Slovakia (Miklós et al. 2006), the representative landscape types of Slovakia (Bezák et al. 2010) and model of representative geoecosystems at the regional level (Izakovičová et al. 2011).

Most of these regions contain several REPGES types from which a representative type is usually one covering the largest area, although if there are more types of relatively same size, more than one representative type can be identified. In the same vein, in some regions, the representative types were not necessarily the most extensive ones, but rather very characteristic parts of the given region.

Such approach seeks to create spatial units with one or more REPGES types, which should maintain geoecodiversity in each region, and therefore it is necessary to **preserve and protect them and include them into the territorial system of ecological stability**. The essential principle of each step is that each region has to have a representative type of REPGES and, on the other hand, each REPGES type in some region functions as representative.

This constitutes a strategic value of the REPGES concept.

4.2.1.3 State of the Protection of Representative Geoecosystems Types

One of the articulated goals of REPGES identification is to elaborate a **scheme for the strategy of geoecodiversity protection**.

In practical terms, it means to design such a scheme that would include a list of all strategically important REPGES types in each territory, which need to be preserved and protected, and analysis of their degree of protection.

The scheme would explain which representative REPGES types are under adequate, inadequate or no protection.

Globally, nature protection in Slovakia is at a high level, as regards the number, surface area and quality of protected areas. Nevertheless, after developing the systemic scheme of REPGES and its subsequent comparison with the protected areas we can conclude that many REPGES remain unprotected. A more detailed analysis of this issue is presented in the Atlas of representative ecosystems of Slovakia (Miklós et al. 2006) and the study representative landscapes of Slovakia (Bezák et al. 2010). In general, there is still prevailing tendency to promote the protection of "attractive" REPGES types (high-mountain types, karstic types) to "ordinary" types. Less protected are the REPGES types of loess tables, and hills, lower highlands, plateau sub-upland hills and promontories, although they are very important from the ecological, area and bioproduction perspective. Regionally, northern and western regions are strongly preferred over southern and eastern regions of Slovakia.

The strategy of nature protection was radically affected by the introduction of the Natura 2000 network in Slovakia.

4.2.1.4 Output of the Representative Geoecosystems Concept for the Territorial System of Ecological Stability

Strategical output of the procedure is an identification and delineation of REPGES types, which are crucial for the preservation of geoecodiversity in a given territory as well as determination of its use in nature protection. These REPGES types should comprise the skeleton of TSES and function as biocentres or core areas (Jongman 1996; Nowicki et al. 1996; Troumbis et al. 1995).

This requires the following two steps.

Step 1 Comparison of real and potential representative geoecosystems, it means, the assessment of the current landscape structure in the territories of selected REPGES for TSES

Such synthesis will yield an image demonstrating whether:

- There is a real landscape element corresponding with the defined representative REPGES type in the concerned region, and therefore it can be included in the skeleton of TSES.
- This also implies
- Which potential REPGES type does not include a real corresponding and ecologically suitable segment in the territory, and therefore the most suitable potential, however currently badly situated, REPGES should be identified and designated.

Step 2 Comparison of selected REPGES for TSES—real and potential—with the legal protection of nature and natural resources—positive socio-economic factors

This synthesis should substantiate whether the REPGES types, to be included in the skeleton of TSES, are under protection or not. Based on this:

- REPGES which are not yet protected are proposed for protection,
- Protected REPGES need to have the degree of their protection reassessed and propose its strengthening,
- The practical output is
- A list and map projection of the areas proposed for the elements of the skeleton of TSES,
- Assessment of their real state,
- Proposal or re-evaluation of their protection level.

4.2.2 Synthesis Indicators of the Properties of Abiotic Landscape Elements—Creating of Abiocomplexes

Abiocomplex (ABK) is a complex time–space, matter–energy, spatially delimited geosystem, consisting of abiotic landscape elements interconnected by reciprocal relations. It is characterised by specific structure of laws, development and functionality. It interacts with flora, fauna, human society and its socio-economic activities and products and creates condition for their existence.

The goal of spatial synthesis of abiocomplexes is to build a properly designed spatial database—map of abiocomplexes—which can be utilised for the purposes of TSES, namely for:

- Description of selected factors and processes threatening ecological stability in the landscape (soil erosion, landslides, water pollution etc.),
- Defining opportunities (resources, potentials and existing restrictions) to utilise a given territory,
- Create proposals for ecologically optimum methods of use and protection of the landscape—proposals for ecostabilising measures.

The output of this step represents a formulation of localisation criteria of TSES, i.e. where and which elements should supplement the existing skeleton of TSES, and where and what kind of ecostabilising measures should be adopted (Miklós 1990).

Abiotic complexes need to be specified at least based on the relief, rocks and soil. Additional information about abiotic landscape elements (waters, air) can be partially acquired from the interpretation of the relief–rock–soil complex. ABKs need to be characterised by the data on relations between the ABK elements, types of processes occurring in the landscape and the patterns of spatial differentiation. Therefore, it is necessary that this job is performed by a sufficiently experienced and qualified professional.

The output of this procedure is a delineation and description of abiocomplexes (ABK) representing a homogeneous spatial unit with a clearly defined set of analytical properties. It is a difficult process, which requires the synthesis of relief–rock–soil units with such a set of abiotic properties, which allow assessing of each type of abiocomplex in terms of its potential susceptibility to selected conditions and processes threatening ecological stability, and develop a proposal for type and location of ecostabilising measures.

The most widely utilised procedure in this respect is a superposition of analytical maps and simultaneous employing of a method of leading factor. Given the inaccuracies of the map materials and extensive use of GIS technologies for superposition of maps, we would like to point out a high probability of results exhibiting illogical combinations of properties, unrealistically small territories, shapes and location of areas. Therefore, it is required to re-evaluate and adjust the synthetic units according to the logical and functional relations within the framework of abiotic complexes and to inspect the problematic areas in the field because only mechanical superposition might not suffice.

The process of creating of abiocomplexes is addressed in many studies discussing LANDEP methodology (e.g. Špinerová 2010).

The abiocomplex model can be formulated as follows:

$$ABK = (R_x, H_x, V_x, P_x,),$$

where

R_x Indicator of relief properties, e.g. morphographic–positional type, geometric form, inclination of relief,

H_x Indicators of surface and groundwater, for example groundwater level, permeability of rocks, chemism,

V_x Indicators of substrate and geological base, e.g. lithotype, chemism, texture,

P_x Indicators of soil properties, e.g. soil type, skeletality of soils, depth.

Each individual abiocomplex area is assigned its code and organised in a table. Abiocomplexes determine the locating preconditions of the TSES elements, especially ecostabilising measures.

4.2.3 Synthesis of Socio-economic Factors Promoting the Functioning of the Territorial System of Ecological Stability

Synthesis of positive socio-economic factors (SEF) is a integrated spatial projection–overlapping–of the areas and borders of nature protection, protection of natural resources, cultural monuments and the elements of CLS, selected on the basis of their biotic characteristics. The mapping units depend on the hierarchical level of TSES, character of the territory and projection scale. The combinations of positive elements limit proposed activities and determine the proposals for TSES elements.

An example of mapping units of SEF, which promote functionality of TSES on local level, is given in Table 4.3a, b.

The presented mapping units can be also expressed synthetically which would then yield spatial complexes of SEF that characterise the implementation conditions for the creation of TSES. The larger area the positive SEF units cover in the given territory, the more favourable conditions for the implementations of TSES they have, and it can be claimed that they are also more ecologically significant from the point of legal safeguarding.

Admittedly, a certain territory will contain only some or none of the aforementioned categories, and sometimes SEF categories condition each other's occurrence. The combinations of positive elements are projected in maps of limiting factors for human activities, and determining the proposal for the TSES elements, they function as the implementation conditions of TSES.

4.2.4 Synthesis of Socio-economic Factors Threatening Functionality of the Territorial System of Ecological Stability

The goal of the spatial synthesis of stress (negative) factors is to create a complex spatial projection of all the stress factors occurring in the concerned territory and gauge the scope of their impact. The complex spatial projection constitutes the territorial system of stress factors (TSSF) which represents a barrier with respect to the elements of the territorial system of ecological stability.

The spatial synthesis of negative factors can be obtained by superimposing of individual SEF onto a single map.

Table 4.3 Mapping units of positive socio-economic factors of local TSES

a) Mapping units of positive socio-economic factors of local TSES	
Elements of primary and secondary landscape structure as base for SEF	SEF defined in various regulations
Specific landscape structure of a predominantly natural character	Different categories of nature conservation areas and their buffer zones: • Protected landscape area (PLA) • National park (NP), NP protection zone (PZ NP) • Protected area (PA), PA protection zone (PZ PA) • Nature reservation (NR) • National nature reservation (NNR), buffer zone NR and NNR • National natural monument (NNM) • Nature monument (NM), caves, natural waterfall • Protection zone of NM and NNM • Private protected areas
	Zones of PLA and NP
	Natura 2000 • Special areas of conservation; SCI—sites of community importance • Special protection areas; PBA—protected bird areas
	Sites of the UNESCO World Cultural and Natural Heritage
	Biospheric reservation of the UNESCO's Man and Biosphere Programme (MAB UNESCO)
	Wetlands by Ramsar Convention
	Elements of TSES in development documents, plans and projects • Protected landscape element—biocentre TSES • Protected landscape element—biocorridor TSES • Protected landscape element—interactive TSES element
Significant landscape structures and elements	Historical landscape structures
Important species and trees	The gene sites of flora and fauna Protected plants, animals Protected trees (PT) and their buffer zones Protected minerals, fossils/petrifications

(continued)

Table 4.3 (continued)

b) Mapping unit of selected socio-economic factors of the protection of natural resources of local TSES	
Elements of primary landscape structure as base for SEF	SEF defined in various regulations
Forests	Protective forests: • In exceptionally unfavourable places • High-mountain forests, forests in the scrub pine belt • Other forests with predominant protection of soil
	Forests of specific designation: • In protective zones of water resources • In protective zones of healing and mineral resources (spa forests) • In nature conservation areas • Forests designated for research • Forests with medical–recreational function, nearby medical institutions • Suburban forests with health–recreational function, forest parks • In recognised animal and pheasant grounds • Forests under the influence of emissions
	Commercial and other forests: • Timber production forestry • Forest nursery • Military forests • Forests land in sites of community importance • Forest land in declared protected bird area
Water	Protected water area (protected area of natural accumulation of water)
	Water supply rivers and reservoirs
	Catchment of water supply rivers and reservoirs
	Important water flows
	Zones of hygiene protection of underground and surface water of first to third degrees (protective zone of water sources)
	Protective zones of reservoirs
	Water suitable for various functions: bathing, irrigation, for fish breeding and reproduction of original species of fish
Water with specific properties (natural healing sources)	Protective zones 1. to 3. level for natural healing water sources

(continued)

Table 4.3 (continued)

b) Mapping unit of selected socio-economic factors of the protection of natural resources of local TSES	
Elements of primary landscape structure as base for SEF	SEF defined in various regulations
	Natural healing sources of gases and emanations
	Natural healing resources from peat fens, sludges and other Earth healing sources
	Protective zones 1. to 3. level for natural sources of mineral water
	Spa cities (natural healing spa), protective zone of spa cities
	Interior spa territory
	Exterior spa territory
Climatic conditions beneficial for healing	Protective zone of climatic condition beneficial for healing
Agricultural soils	Registered parcel, culture, notes on culture
	Bonited soil-ecological units (BSEU)
	Registered permaculture: vineyard, orchards, permanent grassland for grazing, growing areas
	Agricultural production areas
	Investment to land (melioration, irrigation)
	Agricultural land in sites of community importance
	Less-favoured areas in registered soil blocks
Habitat of fish and animal	Protected fishing areas
	Fishing grounds
	Fish-farming flows
	Recognised animal keepers
	Independent pheasantries
Minerals	Protected area of deposits with reserved minerals
	Protected area of deposit of non-reserved minerals
	Natural rock structures for storing gas
Historic structures and sites	Cultural monuments and national cultural monuments
	Memorial City Reservation (MCR) and Folk Architecture Reservation (FAR)
	Monuments zone—city and landscapes
	Archaeological find
	Protective zones of the memorial fund

The mapping units include:

- Primary stress factors, whose negative impact is their mere physical presence in the territory, such as barriers represented by physical structures, which are also the elements of the current landscape structure, and therefore their spatial projection is identical with the structures in the CLS evaluation,
- Secondary stress factors, including either
  - Deterioration factors—types of pollution of the elements of primary and secondary landscape structure which are unstable, but their spatial occurrence is also mapped,
  - Intangible factors, such as various protection, hygienic and safety zones.

The key of the mapping units of stress factors of a local TSES is exemplified in Table 4.4a, b.

The output of the syntheses is demarcation and re-evaluation of the areas with different combination of stress factors.

Currently existing negative factors are selected and analysed for each specific territory. There is no assumption that all of them can occur in the same territory because the occurrence of one element may prevent the occurrence of the other.

These territories with the combinations of stress factors are subsequently assessed individually according to the types and intensity of the impact of the factors, the load levels are determined, and this constitutes **the territorial system of stress factors**.

Based on the combination of the stress factors, typological territories of stress factor combinations can be identified:

- Areas with the multifunctional impact of stress factors,
- Areas with accumulation of various stress factors,
- Areas with monofunctional dominant stress factors,
- Areas with dominant stress factors resulting from industrial development,
- Areas with dominant stress factors resulting from agriculture,
- Areas with dominant stress factors resulting from tourism,
- Areas with dominant stress factors resulting from urban development,
- Areas with dominant stress factors resulting from forest management,
- Areas with dominant stress factors resulting from water management,
- Areas with two-pair or three-pair dominating stress factors.

4.3 Landscape-Ecological Interpretations—Special-Purpose Classifications of Landscape-Ecological Complexes

Landscape-ecological interpretations are considered a key, albeit rather difficult phase, of the TSES methodology. The objective is to evaluate real, measurable and otherwise objectively evaluable properties of the landscape-ecological indicators

Table 4.4 Mapping units of the selected socio-economic factors of residential, production, transport-technical and communal activities and other planning and development documents of local TSES

a) Mapping units of the selected socio-economic factors of residential, production, transport-technical and communal activities and other planning and development documents of local TSES	
Elements of primary and secondary landscape structures as base for SEF	SEF defined in various regulations
Residential areas	Residential zones of individual and complex apartment building
	Zones of public services
	Cultural and school zones
	Zones of relaxation and sport, parks and public greenery
	Protective zones of hospital areas
Recreational areas	Declared recreational area, zones and centres of varying levels
	Declared campgrounds
Production areas	Hygienic protective zones—HPZ/safety zones—SZ of industrial areas areas
	HPZ/SZ of quarries and mining sites
	HPZ/SZ of waste landfills
	HPZ/SZ of wastewater treatment
	HPZ/SZ of agricultural objects
	HPZ/SZ of agrochemical businesses
	HPZ/SZ of animal productions
	HPZ/SZ of store of manure and field airports
Transport and other structures	Protective zones of railways
	Protective zones of roadways
	Protective zones of airports and landing path of airports
	Protective zones of ports
	Protective zones of electrical cables
	Protective zones of gas facilities
	Protective zones of gas and oil pipelines
	Protective zones of cables
	Protective zones of storages
	Military zones

(continued)

Table 4.4 (continued)

b) Mapping units of the selected socio-economic factors of pertaining to the deterioration of the environment of local TSES	
Elements of primary landscape structure as base for SEF	SEF defined in various regulations
Sources of pollution	Stationary sources of atmospheric pollution • Particularly important technological unit ("large source") • Important technological unit, if not part of a large source ("medium source") • Other technological units ("small source")
	Mobile sources of air pollution
	Sources of water pollution (differed according to size)
	Noise sources (depending on size)
	Sources of radiation
Territory with air pollution	Area of air pollution • Area of air quality management • Defined parts of zones or agglomerations with an area of 50 km2, where there occur odorous pollutants in concentrations which are nuisance to the population
	Area burdened by ground-level inversions
	Area burdened by noise (by degree)
Territory with water pollution	Sections of polluted water courses (by category)
	Area of polluted groundwater above the norm
	Sensitive areas
	Vulnerable areas
	Declared inundation area
Soil damage and soil with reduced conditions	Territory with contaminated soil
	Erosion of damaged soil (by degree)
Damaged vegetation	Damage to vegetation (by degree)
	Forests damaged by emissions
Other areas threatened by stress factors	Environmental burden according to register of environmental burdens A, B, C
	Area threatened by landslides
	Area threatened by avalanches
	Undermined area
	Area of increased seismicity
	Area of increased natural radioactivity

and identify such characteristics—interpreted indicators—which could function in the process of elaborating of TSES as criteria and arguments to promote preservation, modifying or proposal for new TSES elements and ecostabilising measures.

The problematic aspect is that these indicators are unmeasurable and only partially quantifiable and, in many instances, are a result of expert assessments and estimations, especially in case of interpretations of biota. Unsubstantiated characteristics and arguments cannot compete with the technical and economical quantified criteria when a dispute arises. Therefore, the scientific course should consider enhancing the methods of landscape-ecological interpretations with sufficiently cogent indicators (Turner and Gardner 1991).

The output of the interpretations is usually the assessment of complex spatial units—abiocomplexes, biocomplexes and landscape-ecological complexes based on the interpreted indicators, which can be also expressed by means of quasi-quantitative value scales (e.g. degree of ecological stability)—meaning by the classification category of the interpreted indicator.

The interpretation procedures can also be very diverse and intricate. The ultimate goal is to produce the most comprehensive representation of the significance of a particular area or a CLS element relevant for the design of TSES, e.g. according to the degree of their vulnerability, significance, rarity, ecological stability and other criteria.

The interpretation procedures are categorised into several thematic and functional blocks:

- *Interpretation of the indicators of properties of abiotic complexes and their subsequent classification* describes the conditions for the application of ecostabilising measures, and vulnerability of geosystems to degradation and eradication of conditions favourable for habitat development—especially spatial differentiation of the functional indicators such as cultivability, accessibility, but also erodibility, waterlogging, salinity, dryness, susceptibility to gravitational deformation, avalanche danger and other.
- *Interpretation of indicators of properties of biotic complexes and the current landscape structure and their subsequent classification*—in most cases the interpretation aims to determine internal ecological stability of CLS elements and their ecostabilising functions; it is based especially on ecological and biological characteristics of CLS.
- *Interpretation of indicators of stress factors and their subsequent classification* determines the degree of threat to ecosystems according to the intensity of their impact.

If there are sufficient materials available, complex special-purpose landscape-ecological classification of the ecological quality of a territory can be conducted—oftentimes referred to as the ecological stability degree of landscape-ecological complexes.

4.3.1 Interpretation of Abiotic Complexes and Their Subsequent Classification

This interpretation should provide the information on location criteria for human activities and ecostabilising measures necessary to ensure proper functionality of the skeleton of TSES. These measures include proposals for optimal use of the territory as well as agroameliorative measures. The interpretations are identical with those described in the methodological procedure of LANDEP and in many other studies (e.g. Ružička and Miklós 1982; Miklós 1990; Miklós et al. 1997; Špinerová 2010); therefore, this publication will not discuss them here.

The indicators of the potential of an abiocomplex to maintain or threaten the ecological stability of the current landscape structure are special interpreted attributes. They mainly focus on a degree of the abiocomplex susceptibility to natural threats, which affect not only the ecological stability of the landscape, but also socio-economic utilisation of the territory, and assess objectively determined indicator values of abiocomplex properties.

The following threats are relevant for the territory of Slovakia:

1. Gravitational deformation,
2. Wind erosion,
3. Sheet, rill and pothole water erosion,
4. Various local processes such as avalanches, windthrows.

The evaluation can be performed by employing a calculation method, semi-quantitative or formalised methods, method of expert judgment, as well as empirical evaluation of the actual combination of factors determining the incidence and intensity of a given phenomenon. The input information draws on the results yielded from the analyses and syntheses of abiocomplexes. Additional information can be also obtained from other sources, e.g. the archive of the D. Štúr State Geological Institute, other institutions as well as field research.

If a sufficient amount of input data is available, relatively simple procedures can be utilised, many of which were published in numerous domestic studies (Tremboš, 1994, Minár and Tremboš 1994, 1995; Pasák et al. 1983 a pod.).

The semi-quantitative methods use order values, weight coefficients and other statistical methods (Tremboš 1998a, b; Tremboš et al. 1997; Midriak 1999). Subsequently, abiocomplexes are categorised into relative value scales according to their potential susceptibility to current natural threats (1. degree—highest susceptibility to 5. degree lowest susceptibility to natural threats).

4.3.2 Interpretation of Biotic Complexes of the Current Landscape Structure and Their Classification

This interpretation is one of the key steps in the development of TSES. Based on the whole-territory mapping of habitats, we evaluate their diversity and favourability of their state. This stems from the maps of current flora and fauna. The assessment of the favourability of habitat state can be conducted with characteristic and indicative taxons using the material produced by the State Nature Conservancy of the Slovak Republic (Polák and Saxa 2005).

The interpretations of biota comprise a wide range of special-purpose classifications, especially concerning the indicators characterised by the terms such as rarity, diversity, stability, resistance, resilience, significance as well as the most used term—ecological stability. In any case, these interpretations aim to find an appropriate criterion to determine the intrinsic value of a CLS element and this could also serve as an argument to specify which CLS elements would suitably perform required functions of biocentres, biocorridors, interactive elements and ecostabilising measures. A wide range of methods then could be used here including the assessment methods of the significance degree of vegetation elements well suited, especially for the local level.

The significance assessment of vegetation elements in the landscape (especially landscape-ecological, biotic, anti-erosion and nature conservation) has a special status as provides important arguments for the landscape vegetation elements, which should function as biocentres, biocorridors and interactive elements of the local TSES. It is essential to identify at least the most important vegetation formations in the territory (potential TSES elements) that can be compared with respect to their ecological–socio-economical evaluation.

The process of assessing significance evaluates the following characteristics.

4.3.2.1 Structural Characteristics of Vegetation

Landscape and ecological diversity for planning and functional arrangements of optimal landscape structure should be specified by preceding analysis of taxonomic diversity (Jurko 1990). Species richness is determined by the number of species given by the Shannon diversity index H (Begon et al. 1997), and it is a function of a relative vegetation cover and the significance of species and logarithmically correlates with the number of species.

In addition to the species richness, it is also important to determine the spectrum of life forms. It is a complex of features determining the appearance (habitus) of plants as the manifestation of their spatial and temporal integration; it is a reflection of habitat conditions, use of territory and the relationships among plant populations. The spectrum of life forms can (in addition to site conditions) also point to a disturbing anthropogenic impact on vegetation, synanthropy—invasion of ruderal annuals, condition the biological stability of vegetation communities, e.g. resistance to compaction, high winds. The most common classification of life forms was proposed by Raunkier (1905 in Jurko 1990). Distribution of species is determined based on the average coverage of all species in all three layers.

The individual life forms are divided by positioning (protection) of the perennial organs (buds, seeds) of plants as adaptation to seasonality (Ellenberg 1974): herbaceous *chamaephytes* (plants with perennial buds between soil level and 30 cm), *hydrophytes* (acquaphyte), *geophytes* (plants with their storage organs underground), *hemicryptophytes* (have tissues protecting them from adverse weather conditions by their position on the soil surface), *nanofanerophytes* presented by bushes and small trees of 0.5–5 m, *phanerophytes* presented by trees taller than 5 m, *therophytes* (short-life annual or biennial species).

Another important structural attribute is their growth strategies, which express adaptation of taxons to the load of adverse conditions (stress) and biomass disturbance (Ellenberg 1974). It is a set of properties, in which evolutionary selection has proved favourable to maintain the existence of certain population, i.e. for its survival and spreading in time and space. In other words, during evolution plants acquired such properties that allow them to exist and spread in certain habitats. The analysis of life strategies indicates the character of habitat as well as the intensity of anthropic impact, and so the assessment is based on the average coverage of individual taxons of all three layers.

We recognise primary and mixed strategies: *c—competitors* (perennial, competitively strong species biomass production intensity, tolerate only low intensity of stress and disturbance), *r—ruderalists* (species with low competitive potential, tolerate high level of disturbance and are adapted to a rapid expansion into unoccupied areas with relatively favourable conditions, their increased occurence indicates disturbance of the environment) with *"s" stress tolerant* (small, slow-growing species, capable of withstanding adverse conditions) *cr—strategists* (species that indicate habitats with extreme nutrient surplus), *cs—competitive strategists* (stress-tolerating species, adapted to moderate stress intensity (moderately productive habitats) without much biomass disturbance), *sr—stress-tolerating ruderal strategists* (species adapted to temperate unproductive habitats (moderate stress) and withstanding moderate biomass disturbance) *csr—competitive-ruderal stress-tolerating strategists* (species, adapted to the habitats where the competition decreases by means of moderate intensity and biomass disturbance).

4.3.2.2 State Variables of Vegetation

Stability expresses natural–anthropogenic permanence of vegetation formations—and their ability to sustain a stable species composition without additional energy in the form of mowing, grazing, removal of seeding, etc. (Ružičková et al. 1990; Maglocký 1993).

Stability of line formations can be evaluated by means of the modified scale of ecological habitat stability proposed by Míchal (1991) with regard to indigenousity, naturalness, representativeness and demand for additional energy to maintain the values of a given habitat:

10. Exceptionally high stability (indigenous habitats without interventions),
9. Very high stability,
8. High to very high stability,
7. High stability,
6. Medium to high stability,
5. Medium stability,
4. Low to medium stability,
3. Low stability,
2. Very low stability,
1. Extremely low stability (habitats with devastated vegetation, with very low diversity).

Occurrence of anthropophytes indicates a divergence from the ideal state represented by potential natural vegetation (Kostrowicki 1982).

Evaluation of synanthropy–naturalness allows for utilising the spectral representation of individual synanthropy–naturalness categories (Marhold and Hindák 1998; Jurko 1990). For example, species coverage is the evaluation criterion for (Halada et al. 1995) proanthropophytes, apophytes, archaeophytes and non-invasive agrophytes.

Naturalness is interrelated with a degree of hemeroby, which determines a degree of human-induced impact, including all intentional and unintentional anthropic influences, which affect the status of habitat and the conditions for life (Kowarik 1990).

Degrees of hemeroby can be defined according to the distribution and representation of terophytes (T) and neophytes (N) (Jurko 1990). In order to determine the landscape-ecological significance, they can be evaluated and classified as follows (the scale can be modified):

10. Extremely high stability—natural, $T < 2\%$,
9. Very high stability—T: 3–5%,
8. High to very high stability—almost natural, T: 6–10%, $N < 3\%$,
7. High stability—T: 11–15%, $N < 7\%$,
6. Medium to high stability—semi-natural, T: 16–20%, N: 8–12%,
5. Medium stability—T: 21–25%, $N <$ 13–18%,
4. Low to moderate stability—cultivated T: 26–35%, N: 19–22%,
3. Low stability—artificial $T > 36\%$, N: 23–35%,
2. Very low stability—$N > 36\%$,
1. Extremely low stability—devastated.

4.3.2.3 Ecosozological Properties of Vegetation

Regional rarity is a representation of particular species or their communities in a region and does not provide a reflection of the whole territory of Slovakia (Špulerová 2004). The evaluation of regional rarity requires compiling, for instance, a list of common (category 3.), uncommon or rather rare species (category 2.) and rare species, e.g. species with monotopic occurrence (category 1.).

Regional rarity can be expressed by a regional rarity index:

$$R_{eg} = \sum k \cdot nir$$

where n is a number of taxons of ith category and k is a coefficient of that category. The coefficient k acquires the following values: $k = 1$ (category 3, common species), $k = 2$ (category 2, uncommon taxons) and $k = 3$ (category 1, rare taxons).

Endangerment is defined by the occurrence of protected, endangered and rare species recorded in the Red List of vascular plants of Slovakia (Feráková et al. 2001). The degree of endangerment is determined by means of the endangerment index:

$$O = \sum k \cdot pi$$

where pi is a representation of taxons of ith category and k is an endangerment coefficient category (CR—critically endangered: $k = 9$, EN—endangered: $k = 5$, VU—vulnerable: $k = 3$, LR—lower risk: $k = 1$).

4.3.2.4 Production Properties of Vegetation

The forage value of above-ground sections of plants is influenced by water content, chemical composition, content of essential oils, toxic substances, vitamins, smell, morphological and anatomical texture, accessibility for animals, digestibility, palatability, resistance to fungal diseases and other (Jurko 1990).

Individual species of vegetation formations are evaluated based on their forage econumber (Jurko 1990). The analysis of forage potential is determined by average coverage of species of particular forage value categories. The forage potential is then calculated as sum of per cent proportion of plant with various econumbers excluding the proportion of inappropriate or toxic species.

However, a mere mechanical comparison is sufficient to estimate melliferous potential because of the proportion of phytomass, i.e. coverage not as significant for pollinators as the actual amount of flowers (Jurko 1990). From the practical perspective, the time of flowering, colour salience as well as the quality of sources (proportion of individual nectar categories, pollen, honeydew) need to be taken into account.

Melliferous potential is expressed by the potential of nectar supply and pollen in a particular vegetation formation. The analysis of melliferousness is derived from the average cover of species and their respective categories. The evaluation can also draw from the database of attributes of species proposed by Jurko (1990).

4.3.2.5 Ecological Significance of Biotic Complexes

(a) **Landscape-ecological significance of vegetation**

Landscape-ecological significance is a complex of interpreted attributes of the landscape, which determine the functioning of ecosystem processes that maintain favourable conditions for preservation, regeneration and recovery of the gene pool, natural resources, ecological stability, biodiversity and other landscape functions (Hrnčiarová 1998).

Landscape-ecological significance can be interpreted as a complex interactive function of vegetation stability, ecological significance and a proportion of rare and endangered species as well as other indicators (Miklós et al. 1986).

The main parameters of a complex assessment of landscape-ecological significance include the vegetation attributes mentioned in the previous paragraphs (Jurko 1990): degree of hemeroby (H), endangerment (O), regional rarity (V_r), diversity (D), forage potential (P_{to}), melliferous potential (P_m) and stability (S_{pa}). Each criterion can be assigned increasing values from 1 to 5. Landscape-ecological significance is calculated as follows:

$$V_{ke} = \frac{(H + O + V_r + D + P_k + P_m) \cdot S_{pa}}{100}$$

The evaluation of landscape-ecological significance of vegetation is essential for the proposal for TSES elements—biocentre, biocorridors and interactive elements, as well as ecostabilising measures.

(b) **Conservation significance of vegetation**

The nature conservation interpretation of vegetation rests on the evaluation of the proportion of rare and endangered species as well as local assessment of rare communities and habitats. Important nature conservation evaluation criteria include species richness, naturalness, endangerment and regional rarity of species. Each criterion can be assigned a value on the scale from 1 (lowest values) to 5 (highest values). The final value represents a total sum of points awarded to the individual characteristics (the higher the conservation significance value, the greater the significance for the skeleton of TSES and the protection of landscape elements).

The assessment of nature conservation significance is taken into account in the proposals for TSES legislation.

(c) **Biotic significance of vegetation (habitat significance)**

Biotic significance can be assessed using the following criteria (Diviaková 2010): Shannon index of diversity of fauna and flora communities, minimal formation width (the area of vegetation formation), type of adjacent habitats, linkage of lines with other formations (interconnectivity), layers of formation. Each criterion can be evaluated on the scale from 1 (lowest values) to 5 (highest values). The final value represents a sum of points awarded to the individual characteristics (the higher the biotic significance value, the greater the significance for the TSES skeleton).

The evaluation of biotic significance of vegetation is essential for the proposal for TSES elements—biocentres, biocorridors and interactive elements, as well as ecostabilising measures.

(d) **Ecostabilising and landscape-ameliorative significance of vegetation**

Evaluation of this significance of vegetation is problematic as almost all ecological and environmental functions are of interactive and multifunctional character. These include the following cumulative functions:

- Soil-protecting (e.g. water erosion, wind erosion, drought prevention),
- Microclimate (e.g. aerodynamic, microthermic, interceptive functions)
- Ameliorative (e.g. retention, rehabilitation, recultivation function)
- Health sanitary (e.g. inhibitive, barrier, filtration function)
- Other (e.g. refuge, fire protection, aesthetic function).

The effectiveness of each function can be objectively measured (e.g. anti-erosion, microclimate, etc.), some allow only approximate estimation, whereas other can be more accurately specified based on the analysis of aforementioned ecological characteristics (biotic and abiotic characteristics).

The anti-erosion function can be relatively simply defined by erosion vulnerability models (Lieskovský 2006; Kočický et al. 2008 and others). First of all, it is necessary to identify relevant erosion types for evaluation: quantitative and qualitative characteristics of biotic components as well as abiotic environment properties that affect erosion are reflected in the erosion models. The scope of the anti-erosion effect can be defined by contrasting potential (simulation of soil loss induced by the absence of vegetation cover and ploughing along a fall line) and a real model (soil loss with protection and inhibitory vegetation impact) of a territory threatened by a particular natural process or by means of comparison of the variant of a current erosion model and the variant which lacks the data on a particular vegetation element, whose significance is to be examined (Diviaková 2010).

The assessment of anti-erosion significance is included in the proposals for ecostabilising measures.

The evaluations of structural, status, ecosozoological characteristics and production attributes of vegetation based on the significance of biotic factors are aptly described in many studies (e.g. Jurko 1990; Špulerová 2004; Diviaková 2008, 2010 a o.).

4.3.2.6 Degree of Ecological Stability of the Secondary Landscape Structure/Land Cover Elements

The concept of the degree of ecological stability of CLS elements is a frequent one, but still relatively insufficiently explicated. The assessment of CLS elements should express their merit with respect to the environment and territorial system of ecological stability. It works on the assumption that a relative degree of ecological stability is inversely proportional to the intensity of anthropic impact on the ecosystem; however, the evaluation of a degree of anthropic influence might not always be entirely objective as it depends on evaluator's own skills. More objective evaluations should observe complex interpretation criteria specified in the foregoing paragraphs (special-purpose interpretations of biotic elements), but simple expert evaluations based on experience are also quite common.

Such evaluation is often used, for example, for designation of the elements of "the skeleton of ecological stability" as argued by the authors of the landscape-ecological school in Brno (Buček et al. 1984), where one criterion is a comparison of the potential and actual state of ecosystem in the landscape. From the perspective of ecological stability, a five-degree scale has been devised for the evaluation of the significance of landscape segment (Löw et al. 1995):

1. No significance (e.g. built-up areas and roads with asphalt or concrete surface),
2. Low significance (e.g. large-block fields, intensive vineyards, hop fields etc.),
3. Medium significance (in terms of the TSES objectives includes intensified meadows, extensive use of permanent crops, etc.),
4. High significance (e.g. extensively used permanent grasslands, mixed forests, etc.),
5. Very high significance (mainly natural forests, natural grass–herbaceous communities, wetlands, peat bogs, watercourses and areas with natural beds and banks and accompanying aquatic and coastal communities).

A scheme for the assessment of CLS elements based on this scale for a local TSES (also a regional TSES) is shown in Table 4.5.

4.3.2.7 Ecological Quality of Spatial Landscape Structure

As most methods employed to determine the degree of ecological stability are underpinned by a consideration that the higher ecological quality a particular element of current landscape structure has, according to biological, ecological, but also other utilitarian functions, their positive spatial impact becomes more prominent; e.g. it has higher water retention capacity, soil erosion and filter capacity, higher safety, hygienic and aesthetic functions (Hellmund and Smith 2013). A similar procedure was adopted for the project of the Environmental Master Plan of the Slovak Socialist Republic in 1986, where the knowledge of ecological quality of CLS elements, acquired through many years of expert experience, was formulated as a coefficient

Table 4.5 Expert evaluation of the degree of ecological stability of the current landscape structure elements of TSES

Degree of stability	Current landscape structure elements
A	*Forest vegetation*
5	Oak forests
5	Hornbeam–oak forests
4–5	Pine–oak forests
5	Birch–oak forests
5	Birch–alder forests
5	Hardwood floodplain forests—hornbeam–ash floodplain forests
5	Transitional floodplain forests—oaks–ash floodplain forests
5	Transitional floodplain forests—elm–ash floodplain forests
5	Softwood floodplain forests—willow–poplar forests
5	Birch–oak forests
5	Oak–beech forests
4	Forest-steppe–pine forests
4–5	Oak–pine forests
3–4	Spruce–pine forests
5	Ash–alder forests
5	Beech forests
5	Fir–oak forests
5	Oak–fir forests
5	Fir–oak forests
3–4	Spruce–pine forests
4–5	Oak–pine forests
5	Beech forests with admixed fir
5	Beech forests with admixed oak
5	Oak–fir forests
5	Mountain beech forests
4–5	Spruce–fir forests
5	Maple–fir forests
4–5	Beech trees with admixed fir
5	Fir–maple forests
4	Pine–spruce forests
4–5	Fir–beech forests
4–5	Fir–beech forests with admixed spruce
4–5	Mountain spruce forests
4–5	Alpine spruce forests
5	Dwarf pine

(continued)

Table 4.5 (continued)

Degree of stability	Current landscape structure elements
B	*Non-forest woody vegetation (NFWV)*
5	Groups of natural NFWV
3	Groups of planted NFWV
5	Riparian stands
43–5	Line natural NFWV
3	Line planted NFWV
C	*Permanent grasslands*
5	Vegetation of stagnant water, bogs and abandoned channels
5	Vegetation of moors and mires
4–5	Wet lowland meadows
5	Wet foothill and mountain meadows
4	Extensively used meadows and pastures
3	Intensively used meadows
2–3	Intensively used pastures
3	Dry meadows and pastures
5	Subalpine vegetation
5	Alpine vegetation
D	*Arable land and permanent crops*
2	Large-block arable land (fields)
2–3	Large-block arable land with low occurence of NFWV
3	Large-block arable land with occurence of NFWV
3	Small-block arable land—narrow-stripped fields
3	Small-block arable land—terraced fields
2	Large-block vineyards
2	Terraced vineyards
3–4	Small-block vineyards
3–4	Abandoned vineyards
3	Orchards and plantations
3–4	Abandoned orchards
2	Hop fields
2–3	Gardens
3	Nurseries for ornamental and fruit trees
3	Uncultivated agricultural areas (fallow lands)
E	*Mosaic structures*
3–4	Mosaic of small-block arable land, meadows and NFWV
4	Mosaic of meadows and NFWV

(continued)

Table 4.5 (continued)

Degree of stability	Current landscape structure elements
F	*Watercourses and water areas*
5	Watercourses natural (unregulated)
2–3	Watercourses regulated
5	Oxbow lakes and meanders
2–3	Channels
5	Natural water surfaces (lakes, mountain lakes)
3	Dams
3–4	Abandoned gravel pits
3–4	Ponds
G	*Elements without permanent vegetation cover, or at its initial stages*
4–5	Natural rock formations
3–4	Natural sand and gravel deposits
3	Quarries
2–3	Gravel pits, sandpits, clay pits
3	Overgrowing quarries
3	Overgrowing sandpits, clay pits
2–3	Other areas without permanent vegetation cover, or at its initial stages
H	*Industrial and mining areas*
1	Factory premises (shop floors, warehouses, a.o.)
1	Structures of surface mining structures
1	Structures of subsurface and underground
1	Industrial waste dumps
1	Waste ponds
I	*Pipelines*
1	Extra high-voltage power lines
1	High-voltage power lines
1	Medium- and low-voltage power lines
1	Pipelines
J	*Traffic elements*
1	Highways
1	I. class roads
1	II. and III. class roads
1–2	Unpaved roads
1	Railways
1	Cog railway
1	Railway areas
1	Parking areas
1	Airports

(continued)

Table 4.5 (continued)

Degree of stability	Current landscape structure elements
K	*Agricultural areas*
1	Agricultural production areas
1–2	Field airports
1	Field dunghills
L	*Residential and recreational areas*
1	Residential areas with low-rise buildings
1	Residential areas with high-rise buildings
2	House gardens
1–2	Civic amenities
3–4	Parks
3	Cemeteries
1	Recreational and sports facilities and areas
1	Cottage and log cabin villages
1–2	Caravan parks and campgrounds
1	Spa areas
1–2	Hiking trails
M	*Other areas*
1	Military facilities and areas
1	Waste dumps
1	Building sites
1	Non-used land

(Jurko 1990; Miklós 1986), which modifies the areas of individual CLS elements as indicated by the formula:

$$\boldsymbol{K}_{\mathbf{es}} = \left(\sum \boldsymbol{p}_i \cdot \boldsymbol{k}_{\mathbf{SKSi}}\right) \cdot \boldsymbol{P}^{-1}$$

K_{es}—the coefficient of ecological stability of the territory, p_i— the area of an *i*th CLS element, K_{SKSi}—the coefficient of ecological quality of *i*th CLS element, P^{-1}—the total area of a studied territory.

The final value of ecological quality of an examined territory is represented by a weighted ratio of an ecological quality area and a total area of the examined territory.

Determining of K_{es} is illustrated by an example of Ilijsky stream (Table 4.6, and Fig. 4.1).

This basic result can be further modified and interpreted in various ways; e.g. the change of ecological stability for different periods can be compared, the area of ecological quality with climate comfort and environmental state can be modified, and the ecological quality values of geomorphological regions, protected areas, urbanised areas, etc., can be calculated.

Table 4.6 Coefficient of ecological quality and run-off coefficient of the current landscape structure (CLS) elements in a model territory in the catchment area of the Ilijsky stream

i	CLS element	Area in m^2	% of the entire area	k_{SKSi}	k_{odi}
1.	Deciduous forest	2,260,875	14.5	1	0.03
2.	Mixed forest with the predominance of deciduous trees	1,010,146	25.4	0.95	0.03
3.	Mixed forest with a balanced composition	3,966,003	6.5	0.9	0.04
4.	Mixed forest with the predominance of deciduous trees	215,643	1.4	0.85	0.05
5.	Coniferous forest	244,187	1.6	0.75	0.06
6.	Continuous shrub stands	302,152	1.9	0.7	0.1
7.	Water areas	3540	0.02	0.6	1
8.	Permanent grassland with shrubs	2,101,958	13.5	0.7	0.13
9.	Meadows and pastures	1,091,530	7.0	0.65	0.15
10.	Orchards	334,157	2.1	0.5	0.13
11.	Arable land	3,646,209	23.4	0.25	0.25
12.	Greenery in residential areas	97,180	0.6	0.4	0.13
13.	Sports areas	1236	0.01	0.25	0.2
14.	Buildings and other technical structures	52,384	0.3	0	1
15.	Other residential areas, courtyards	157,479	1.0	0.1	0.7
16.	Stockyard, waste dump	4785	0.03	0.1	0.5
17.	Paved roads and car parks	109,853	0.7	0	1
18.	Rocks, cliffs, furrows	1738	0.04	0.2	0.8

k_{SKSi}—coefficient of ecological stability of an **i**th CLS element
k_{odi}—run-off coefficient index (coefficient of run-off retention for erosion models)

A popular and practically usable interpretation of this indicator is the share of ecologically quality area in m^2 per capita in different administrative units according to basic formula

$$EQA_{\mathbf{inhab}} = \left(\sum p_i \cdot k_{\mathbf{SKSi}}\right) \cdot \mathbf{inhab}^{-1}$$

where

$EQA_{\mathbf{inhab}}$	The ecologically quality area of concerned territory
p_i	The area of *i*th CLS element in the concerned territorial unit (municipality, district, region, district, subregion, microregion)
$k_{\mathbf{SKSi}}$	The coefficient of ecological quality of *i*th CLS element
inhab	Number of inhabitant in the concerned territorial unit

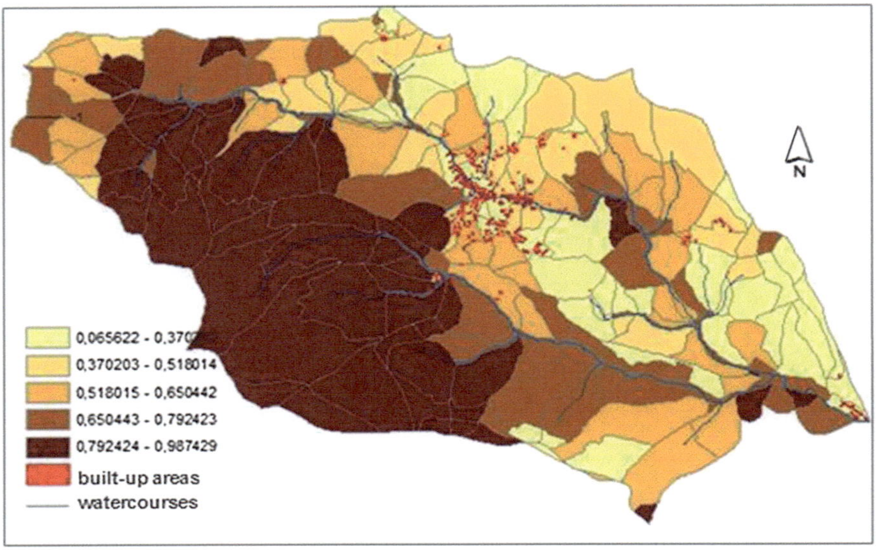

Fig. 4.1 Coefficient of ecological quality in the microcatchment area of the Ilijsky stream

Figure 4.2 is an illustration of the ecologically quality area per capita in the municipality area in the Ipel' River catchment and Fig. 4.3 the ecologically quality area per capita in the natural-settlement subregions and microregions of Slovakia.

The results of the assessment of ecological quality show the relative differences between different territorial units. They can be used for quantitative estimation of the needed area of ecologically stable CLS elements to be completed to territorial unit in order to achieve a desired standard value of this indicator.

4.3.2.8 Ecologically Significant Landscape Segments

The criterion of ecological stability forms the basis for designation of ecologically significant landscape elements. Ecologically significant landscape elements (EVSK) are the segments of landscape consisting of the ecosystems with relatively greater ecological stability where such ecosystems prevail in the area (Wascher and Jongman 2000). They are characterised by the permanence of biota and ecological conditions, which allow the existence of species that constitute the natural gene pool of the landscape. The integration of landscape segments into the skeleton of ecological stability employs *the relative selection principle*. Only the segments with the relatively highest value of biodiversity conservation, and thus the greatest internal ecological stability, are selected. This means that even the segments with a lower degree of ecological stability can be a part of the skeleton of ecological stability in an extensively anthropogenically altered agro-industrial landscape. However, it applies that the skeleton

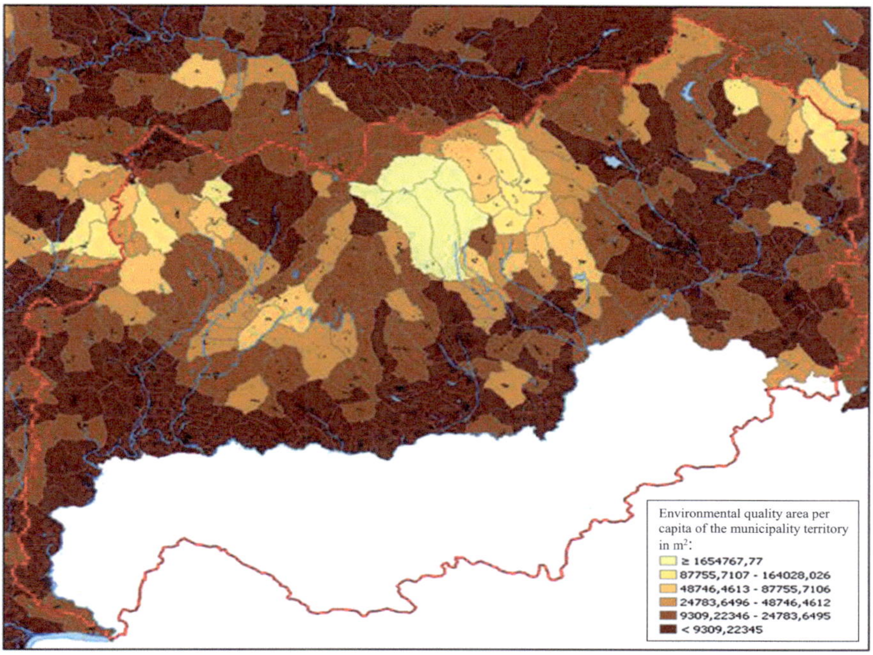

Fig. 4.2 Environmental quality area per capita in the territories of municipalities (Ipel' River catchment area)

should consist of the territories with a higher degree of ecological stability 3–5, the territories from the three highest categories of ecological stability.

When defining the skeleton of ecological stability, EVSK can be further divided by spatial-structural criteria (size and shape, state of current biocenosis). Although the criteria are not entirely objective, their parameters result from expert evaluations. The system of such criteria was proposed, for example, by Buček and Lacina (1984); Löw et al. (1995) include the following characteristics:

- Ecologically significant landscape elements—small territories (1–10 ha) with homogeneous ecological conditions, for example a group of trees, a solitaire in agricultural landscape, natural meadow in the middle of intensively used, artificially fertilised meadows and pastures,
- Ecologically significant landscape units—larger territories (10–1000 ha), where a diversity of environmental conditions allows the existence of several types of communities, for example valleys of upper and middle sections of rivers with forest, rock and wetland communities. Various significant landscape elements can be delineated within landscape units as well,
- Ecologically significant landscape areas are extensive territories (usually more than 1000 ha) with diverse ecological conditions and communities with a high distribution of natural communities. They include landscape protected areas, but also

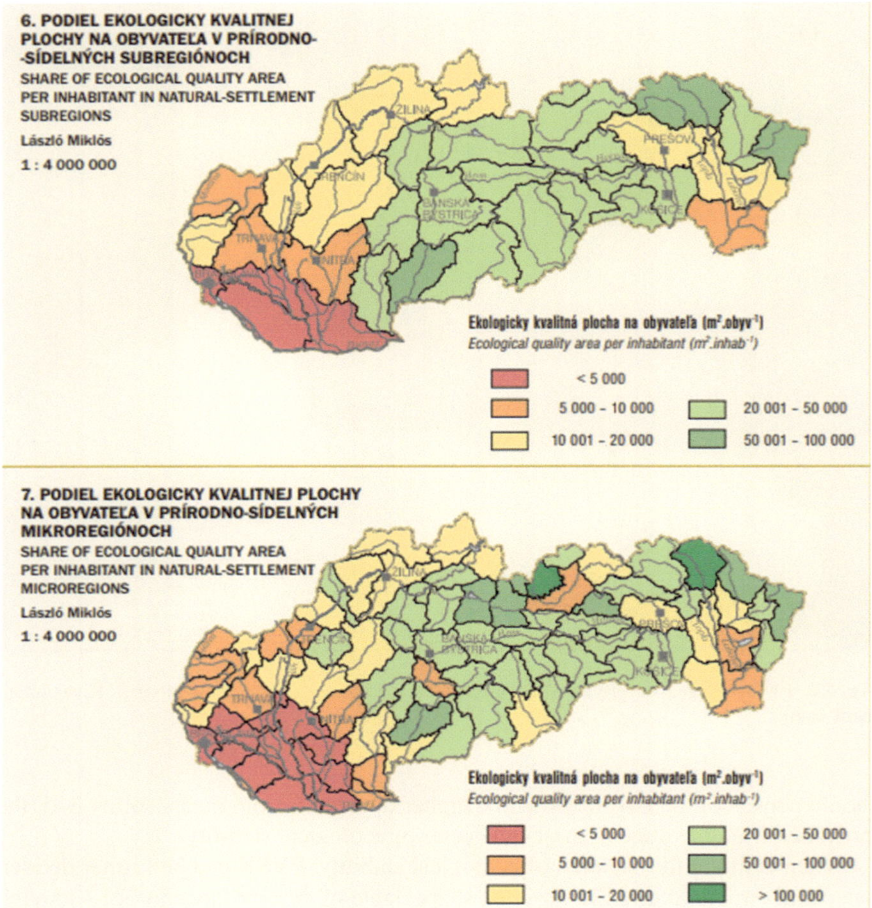

Fig. 4.3 Environmental quality area per capita in the natural-settlement subregions and microregions of Slovakia (Miklós 2002a, b). The legend of maps in Slovak and English

other large areas with predominance of natural forests and species-rich meadow communities. It is advisable to identify important landscape elements and linear communities within ecologically significant landscape elements,

- Ecologically significant line communities have an elongated shape with the predominance of transitional peripheral biocenoses (ecotones). The densest network of line communities in a cultural landscape is formed by watercourses and riparian vegetation. The major line communities also include alleys and treelines comprised of native woody species.

Based on expert evaluations, the EVSK parameters were defined as follows:

- Area and spatial parameters,
  - Surface area: categories up to 10, 10–50, 50–100 ha, over 100 ha,
  - Shape: square, rectangle, circle, line (length, width),
  - Location: in the centre of a bioregion, in the contact zone of bioregions,
  - Distance from the nearest similar habitat:
  - Isolated natural element (within a distance of 2 km no similar element is situated),
  - Similar habitat (same size, smaller, larger) situated in the vicinity (within 2 km),
  - More similar habitats (same size, smaller, larger) are located in the vicinity,
  - Connectivity—(connectivity of line stands, expressed as the number and length of gaps),
  - Character of the environment.
- Representation of habitat types,
- Current state of habitats (satisfactory, highly satisfactory, partially satisfactory, unsatisfactory),
- Species composition, autochthonity of stands,
  - Occurrence of protected and endangered biota species,
  - Occurrence of invasive plant species, the occurrence of quarantine species,
  - Occurrence of introduced tree species,
  - Occurrence of introduced fruit tree species.
- Age composition of stands (forest stands and non-forest woody vegetation),
- Effect of stress factors,
- Genesis of the area,
  - Fragment of indigenous plant and animal communities,
  - Regenerating natural element—a disturbing factor ceased and succession takes place,
  - Cultural-historical origin.
- Ecostabilising significance—(comparing of a current status with potential natural vegetation in forest communities) an ecological stability degree (1–5).

The outcome of the classification should be a hierarchy of ecologically significant landscape segments, which will be further reassessed from the point of their integration into the TSES skeleton.

Slovak legislation partly stipulates the protection of significant landscape elements by the provisions of Act No. 543/2002 Coll. on nature and landscape protection and requires the protection of biocentres, biocorridors and interactive elements, although they did not have the status of a protected area in the past. It is, however, desirable to ensure similar legal protection also for important landscape elements not yet incorporated in the TSES, but threatened by construction, disintegration, isolation, degradation and destruction. The basis for their registration should be provided by *identification cards of ecologically important landscape elements.*

4.3.3 Interpretation of Socio-economic Factors of the Character of Stress Factors and Their Special-Purpose Classifications

The aim of this interpretation and subsequent classification is the identification of the degree of load of socio-economic factors with the character of stress factors. It is necessary to assess the impact of each SEF separately as well as the degree of the complex load caused by stress factors of primary or secondary discrete, line and patch type. The output is the determination of various degrees and combination of current load on the territory (Kočická 2000, 2007).

In the TSES project, the load of stress factors is spatially expressed by the territorial system of stress factors TSSF. The frame of TSSF comprises of cores, lines—corridors and large areas of TSSF.

The cores of TSSF are discrete or small-scale structures, where stress factors accumulate and linear elements meet or intersect. The cores may be surrounded by large-scale areas of SEF with negative impact, which are directly linked with the cores, for example arable land around the farm technical areas, areas of air pollution, soil contamination around industrial buildings, water contamination around landfills.

At the local level, the TSSF cores are typically represented by a discrete or small-scale structures such as industrial areas, sewage treatment plants, storage areas and other technical works, landfills and settling pits, mining areas, traffic areas, agrotechnical buildings and areas, livestock production farms, residential areas of private house or housing estate character, administrative and commercial areas, recreational and sports areas, cottage settlements, garden allotments.

At the regional level, the TSSF cores cover the areas of entire residential units, larger areas of industrial and technical structures, which can be clearly spatially identified and mapped.

Essentially, the special-purpose classification of cores is based on the evaluation of the intensity of their stress impact. As for the primary influence, the main classification criterion is the size of the area and the intensity of their barrier effect. The secondary influence applies the classification criterion expressed by the type and intensity of secondary influence—production of emissions, noise, dust, bacteriological source of pollution, etc. The combination of these criteria allows the classification of TSSF cores into relative scales from the most to the least loaded.

Linear elements—TSSF corridors—represent anthropogenic lines—roads, railways, power lines, pipelines or adversely altered and affected natural linear elements—polluted watercourses, canals. The evaluation of these stress factors should address each stress factor individually, and in case of multiple parallel lines, their barrier and deteriorative cumulative effect must be taken into consideration. The approach adopted for the assessment of individual lines is fundamentally the same as for the classification of the TSSF cores: the surface area and intensity of secondary influence are assessed. Watercourses are evaluated based on the pollution level of surface water, transport corridors on the intensity of transport, power lines and other pipelines on their installed capacity or the width of their protection zones. The result indicates a degree of stress impact of lines on the landscape or a degree of the load of TSSF lines.

Large-scale areas of TSSF are mostly linked with the cores and typically form zones with a cumulative effect of various stress factors around the individual cores. This regards the accumulation of load on the territory resulting from secondary stress factors. The assessment should be conducted based on the intensity of the stress factor present in the evaluated territory. An array of methods can be employed for that purpose. We can use formalised, semi-quantitative methods, for example, to calculate the intensity of secondary stressors in a particular territory or to calculate their weighted sum, while their weight coefficients are determined according to the relative importance of a stress factor.

The results can be categorised into relative scales and classify the examined territories into groups from slightly loaded—without or only local occurrence of stress factors to extremely loaded—a territory with the values of stress factors exceeding the acceptable limits.

4.3.4 Summary of the Special-Purpose Classification of Landscape-Ecological Quality of Current Landscape Structure Elements

The objective of a complex special-purpose classification of overall landscape-ecological quality of territories is to

- Determine the ecological significance of areas,
- Identify their vulnerability to stress factors.

The result can be obtained from a spatial synthesis of partial interpretations and subsequent classifications presented in the preceding paragraphs, specifically:

- *Interpretations and subsequent classification of abiotic complexes,*
- *Interpretations and a subsequent classification of CLS elements,*
- *Interpretations and subsequent classification of stress factors.*

An effective method for creating a general classification of a territory is, for example, a two-level method of spatial synthesis:

1. The first level includes a spatial synthesis of the results yielded from a special-purpose classification of abiocomplexes and special-purpose classification of CLS elements. The outcome represents the classes of ecological quality of CLS

Table 4.7 Example of a classification matrix I

Threats by abiocomplexes	Degree of ecological stability of CLS elements				
	I. degree	II. degree	III. degree	IV. degree	V. degree
1. Category	1	1	2	3	4
1. Category	1	2	3	4	4
3. Category	1	2	3	4	5
4. Category	2	3	4	5	5
5. Category	2	3	4	5	5

The number of classes may vary

Table 4.8 Example of a classification matrix II

Classification I. from Table 4.7	Load by stress factors				
	I. degree	II. degree	III. degree	IV. degree	V. degree
1. Category	1	1	2	3	4
1. Category	1	2	3	4	4
3. Category	1	2	3	4	5
4. Category	2	3	4	5	5
5. Category	2	3	4	5	5

The resulting combinations can be verbally represented as follows: 1—landscape with very low ecological stability, 2—landscape with low ecological stability, 3—landscape with medium ecological stability, 4—landscape with high ecological stability, 5—landscape with very high ecological stability

affected by abiotic threats. These results can be obtained by combining the input classes and special-purpose classifications into a matrix (Table 4.7).

2. The second level includes a spatial synthesis of CLS elements arranged in a matrix in Table 4.7 and the degrees of load of stress factors in a territory (Table 4.8).

The overall classification is a characterisation of the overall landscape-ecological quality of the areas.

4.4 Landscape-Ecological Evaluations

The aim of landscape-ecological evaluations is to determine the capacity of individual CLS elements to perform some of the TSES functions, according to the overall landscape-ecological quality of these elements determined by the procedures described above.

Selected evaluation procedures include the following.

4.4.1 *Evaluation of the Current Ecological Quality of Potential Representative Geoecosystems and Selected Elements from the Point of View of Their Functions Within the Territorial System of Ecological Stability*

This step involves the confrontation of REPGES and the ecological quality of CLS in their territories. The step also requires comparison of maps and reveals:

- If there is a real landscape element with suitable ecological quality corresponding to a defined representative REPGES type in the region, and it is sufficient to include them in the skeleton of TSES. This also implies
- Which potential REPGES types do not have a real corresponding and ecologically suitable segment, and for such REPGES a suitable potential area should be identified and delineated.

The output of this step is

- Proposal to integrate existing, suitable landscape segments into the TSES skeleton,
- Proposal to establish new landscape segments of the TSES skeleton (new CLS elements in the landscape to complement a functional TSES skeleton).
- This provides a fundamental basis for preparing the proposal for a complete TSES skeleton of real and potential TSES elements.

The aforementioned procedure requires a proper application of the assessment criteria. Contents of the chapter on landscape ecological interpretations (Sect. 4.3) indicates that there is an array of sophisticated evaluation criteria . Their application, however, lies in the question—whether an examined geoecosystem or simply studied CLS elements are suitable or unsuitable for their incorporation into the skeleton of TSES—which renders it often subjective and uncertain. Thus far, no coherent manual outlining the application of the assessment criteria has been compiled, so projection teams approach this step differently.

Nevertheless, the following paragraphs will present several, more or less known and declarative groups of criteria.

Biocentres and biocorridors should satisfy the habitat suitability criterion regarding environmental requirements of different groups of organisms in relation to the hierarchical level of TSES. It is necessary to judge which types of biotic complexes should be assessed for the regional TSES (RTSES) and local TSES (LTSES).

Furthermore, it is important to evaluate the vulnerability of selected elements. It is generally recognised that the lower the degree of human impact on the landscape segment is (i.e. its ecological stability is higher), the more suitable the segment is to function as a biocentre of regional and higher significance.

When evaluating the ecological state of CLS with respect to their suitability for biocentres, the following criteria should be observed (according to Izakovičová et al. 2000):

- General criterion: biocentres comprising TSES should meet not only the ecological requirements of the whole complex of plant and animal species, but also those of the vulnerable and protected taxons,
- Representativeness—biocentres represent a whole range of habitats typical for each biogeographical unit,
- Uniqueness of landscape elements,
- Quality of habitats—high representation of natural ecosystems,
- High degree of biodiversity (protection of areas with high genetic, species and ecosystem diversity),
- Occurrence of endemic or critically endangered species (protection of endemic, endangered, rare and retreating species),
- Current state of the landscape,
- Historical development of the territory,
- Importance for migration or distribution of species,
- Surface, spatial and temporal parameters,
- Social restrictions and intentions.

The fundamental criteria for establishing biocorridors include:

- Size of the areas to be linked,
- Distance between equivalent habitat types,
- Character of biocorridors, their width, the presence of barriers,
- Pressure on biocorridor (urbanisation, agriculture, etc.),
- Degree of biocorridor degradation.

The proposals for biocorridors should also reflect the following requirements of (Smith and Hellmund 1993):

- Interconnection of isolated areas, which were combined prior to their settlement, with similar types of habitats. Such areas may include protected areas of various types threatened by isolation,
- Identification of existing migration routes or movement corridors (e.g. riparian growth functioning as traditional corridors for wildlife species),
- Reinforcing interconnectivity of habitats (e.g. old forest stands) harbouring the species sensitive to fragmentation with regard to limited dispersion opportunities or other factors. On the other hand, minimising interconnectivity of artificially disturbed habitats (e.g. overgrown tracks). In the same vein, it is not suitable to connect peripheral habitats overgrown with weeds and large areas of natural habitats,
- Designation of wide corridors along high-level or length gradients in a way which would allow long-distance migration of target species,
- Avoiding long corridors with a lack of suitable local biocentres if the corridor is not wide enough,
- Inclusion of the whole range of habitats (e.g. in a topographic gradient from a river to a mountain peak). If this is not feasible, then all types of habitats should be represented in the whole ecological network,

Table 4.9 Degree of suitability of land ecosystems for regional (and higher) biocentres based on the naturalness of vegetation (Míchal et al. 1991)

Degree of suitability	Relating to		
	Forest land	Agricultural land	Other areas
1	Natural stands (mostly the most preserved "strict" reserves of inaccessible protected forests)	–	Subalpine meadow communities, preserved peat bogs and wetlands of all types including selected spring areas
2	Near-natural forest, regardless of the intensity of use including overmature tree remnants without regeneration	Extensive meadows and pastures with natural species, with the occurrence of protected or threatened plant species, non-fertilised (often categorised as uncultivated land)	Peat bogs and fens relatively unaffected by former extraction, rushes, rocky and sandy communities without noticeable signs of agricultural exploitation
3	Artificially planted stands with a strong admixture of indigenous woody species or monocultures of indigenous ecotypes	Intensively used meadows and pastures with a large representation of natural plant species unaffected by large-scale forms of use (cutting, additional fertilisation, drainage)	Old nature parks, windbreaks, heathlands, fallow farmland with autochthonous animal and plant species

- Avoiding roads or other potential barriers to movement of animals. If possible, include important areas without roads in the ecological network,
- Creation of a suitable CLS by means of indigenous NFWV, so it could function as a corridor and prevent dispersion of species not using linear structures,
- Designing a multicorridor network, so that sufficient reserves and multiple movements are secured. Such a network will be of greater importance, especially in an area disturbed by wind or fire.

Other criteria for biocentres and biocorridors:

The process of the selection of regional biocentres focuses on the most preserved ecosystem types, assessed based on species combination and the structure of vegetation communities and opportunities of its further development (Míchal et al. 1991; Table 4.9).

Another group of criteria regards dimensions of the TSES elements. These criteria were outlined in the early development of the TSES concept by landscape-ecological school in Brno (Löw et al. 1984; 1995; Míchal et al. 1991). Some of them are introduced below:

In addition, supra-regional and higher-level biocentres have different internal organisation. Each biocentre has a delineated biocentre core, which needs to be strictly protected by means of a protection (buffer) zone. Together, they form a

Table 4.10 Territorial demands of different types of organisms (Míchal et al. 1991)

Type of organism (body size)	Subgroup	Minimum area for a viable population (ha)
Soil microfauna (up to 0.3 mm)	–	1
Soil mesofauna (from 0.3 to 1 mm)	–	1–5
Macrofauna of invertebrates (1–10 mm)	–	5–10
Macrofauna of invertebrates (10–50 mm)	Weakly mobile species	5–10
	Mobile species	10–20
	Flying species	50–100
Megafauna of vertebrates	Small mammals	10–20
	Reptiles, amphibians, small birds	20–100
	Large birds and mammals	100–10,000

complete biocentre area. The biocentre cores should be the best-preserved natural communities.

Territorial demands of individual species, given in Table 4.10, are differentiated by their size categories, whereby the ability of minimal populations to survive excludes their permanent existence within the presented territories in complete isolation.

Minimum size of the TSES elements (Table 4.11) is given by the requirements for providing sustained existential conditions, so that they can genetically "replenish" the surrounding territory. The minimal dimension of a TSES element (surface area of biocentre, length and width of biocorridor) is a boundary value, and any dimensions lower than this value render the elements unable to perform their function.

The apparent paradox that the permissible length of a regional biocorridor is shorter than the local one should be reconciled by so-called combined biocorridor. Smaller local biocentres are designated every 400–1000 m (as a permissible length of a simple biocorridor allows) into a regional biocorridor of disproportional length. This substantially extends the length of a functional regional biocorridor.

Minimum required size of a biocentre will vary depending on different species and communities of living organisms as well as the size, ecological conditions and the location of surrounding biocentres.

Table 4.11 Minimum required parameters of biocentres and biocorridors (Löw et al. 1984, 1995)

Biocentre for forest communities in forest vegetation zone (FVZ):	Local (in ha)	Regional (in ha)
FVZ No. 4. beech and FVZ No 5. fir–beech (ha)		20
FVZ No. 3. oak–beech and spruce–fir–beech (ha)		30
FVZ No. 2. beech–oak, FVZ No. 7. spruce and FVZ No. 8. dwarf pine (ha)	3	40
FVZ No 1. oak and floodplain forest (ha)		50
Biocentres for a quatic communities		
running water	more than 100 m	1–20 km
stagnant water (ha)	1	100
Biocentres for communities of wetlands in floodplains		
in 1. oak to 4. beech FVZ (ha)		10
in 5. fir–beech FVZ (ha)		20
in 6. spruce–fir FVZ (ha)	3	30
in 7. spruce and 8. dwarf pine FVZ (ha)		50
Biocentres for meadow communities		
in floodplains (ha)	3	50
other (ha)		30
Biocentres for steppe communities (ha)	3	20
Biocentres for rocky communities (ha)	0.5	10
Biocorridors	Local (m)	Regional (m)
Maximum permissible length of a simple biocorridor for		
Forest communities		
Wetlands	2000	700
Meadow communities in floodplains	2000	1000
in 5. fir–beech to 8. dwarf pine	2000	700
in 1. oak to 4. beech FVZ	1000	400
Steppe in		
1. oak and 2. beech–oak FVZ	2000	400
3. oak–beech and 4. beech FVZ	2000	700
Minimum required width of a simple corridor		
Forest communities	15	40
Wetlands	20	40
Meadow communities	20	50
Steppe communities	10	20

4.4.2 Evaluation of a Degree of Protection of Selected Elements of the Territorial System of Ecological Stability

This step involves a confrontation of proposed elements of the TSES skeleton with positive SEF.

The comparison of areas suitable for their inclusion into the skeleton of TSES, performed in the previous step, provides an overview of the landscape-ecological significant areas suitable for TSES. However, their level of legal protection varies or they are without any protection whatsoever. Conversely, the process can also expose areas of lower landscape-ecological significance, currently endangered, disturbed, devastated, etc., which are already legislatively protected.

The output of this step is

- Proposal to provide territorial protection for new landscape segments,
- Proposal to strengthen the legal protection in certain territories, e.g. protection zones of forest or water sources,
- Proposal to repeal legal protection in selected territories.

4.4.3 Evaluation of a Degree of Threat to the Skeleton of Territorial System of Ecological Stability by Stress Factors

This step includes a confrontation of positive and negative SEF. In practice it means the comparison of the maps of SEF beneficial for TSES and the maps of SEF which threatens the TSES, in other words, the maps of the territorial system of stress factors TSSF and the maps of TSES. The result is a spatial projection of environmental problems, which expresses the threat to landscape by primary or secondary stress factors, but also the projection of physical barriers for TSES elements.

The output is defining three fundamental categories of environmental problems in a territory and determining the degree of threat to the skeleton of TSES as well as physical barriers to TSES elements.

- *Problems related to the threat of TSES elements* arise from stress factors impacting particular real and potential TSES elements. These elements are threatened by physical barrier effect, pressure of physical seizure of land and devastation of TSES elements as well as secondary stress factors—air pollution, water and soil contamination, disturbed vegetation, increased noise levels, light effects, etc.
- *Problems of threats to spatial landscape stability* arise from the impact of stress factors on the elements with stronger ecostabilising effect such as forest and water areas, permanent grasslands, areas of public greenery. These ecostabilising landscape elements are the most frequently threatened by spatial land seizure mostly for housing development, then industrial and transportation pollutants,

chemicalisation in agriculture and forestry, intensive tourism, mining and erosion-accumulation processes, etc.

- *Problems regarding threats to water sources* arise from a territorial collision of stress factors and natural and cultural-historical sources. Such conflict may impair the quality and yield of water sources. The most serious problem concerns the threat to forest sources (seizure of land, damage inflicted by pollutants, pest outbreaks, avalanches, landslides, etc.), threat to aquatic resources (discharge of sewage, leaching from agricultural chemicalisation, etc.), soil sources (erosion-accumulative processes, soil density, soil contamination, waterlogging, drought, soil infection, etc.), cultural and historical resources (decline of historical land-use forms, etc.).
- *Problems of environmental vulnerability* arise from exposure of man and his environment to stress factors. The most significant problems are the heavily polluted air in densely populated areas, intensive traffic load directed through residential formations and consequent exceeding of permissible noise levels, adverse sanitary impacts of livestock production, lack of high-quality water resources to satisfy drinking water demands, that is.

The intensity of these environmental problems is determined by several factors, especially: the kind of stressor, i.e. which stressor is at work; length, i.e. the duration of the impact of the stressor and its intensity, i.e. how intensive the action of the stressor is.

A special group of problems of environmental characteristics relates to the delineation of problematic areas of CLS from the perspective of the threat to abiocomplexes. This can be achieved by a confrontation of maps of special-purpose classification of abiocomplexes with a map of CLS, which facilitates a delineation of problematic areas in which CLS does not comply with the abiotic conditions, i.e. where the current utilisation of territory intensifies the action of natural risk factors.

4.5 Proposals

Proposals represent the final phase of TSES designing. They comprise a whole system of proposals to constitute a fully functional TSES. The TSES proposals can be divided as follows: proposals aimed at the creation of the skeleton of TSES, proposals for landscape and nature protection, proposals for ecostabilising measures, proposals for elimination of stress factors (Miklós et al. 2018).

4.5.1 Proposals for the Creation of the Skeleton of the Territorial System of Ecological Stability

Proposals aimed at the creation of the TSES skeleton can be categorised into the following blocks:

(a) *Proposals for the inclusion of real CLS elements, able to perform the functions of the TSES skeleton, into the system of biocentres, biocorridors, buffer zones and interactive elements.*

This step is underpinned by the procedure described in Sect. 4.4.1, and it requires only the projection of results into the form of a proposal. When designating the real elements of the TSES skeleton, it is crucial to focus primarily on the elements with a high landscape-ecological value. An important part of the skeleton of the lower hierarchical level of TSES is also the re-evaluation of the elements forming the skeleton of a higher hierarchical-level TSES and their representation in the proposal.

(b) *Proposals for designation of new elements of TSES and completion of the skeleton.*

The basis for this step is also the process described in Sect. 4.4.1. The proposal considers a designation of a new TSES element where, based on the potential conditions of an abiotic complex, a representative geoecosystem could exist, but currently has non-existent, or unsuitable vegetation cover and utilisation of the territory. This group of proposals aims to supplement such TSES elements, which could gradually develop into a fully effective representative geoecosystem, to ensure that all types of representative geoecosystems occur in a given territory and each representative geoecosystem is represented by at least one biocentre in the territory.

However, more probable is the proposal to create new TSES elements to supplement missing elements of biocorridors. The implementation of such a proposal lies in planting such vegetation cover whose species composition would correspond with the site conditions of the concerned locality.

(c) *Proposal for reconstruction and revitalisation of TSES elements.*

The step involves re-evaluation of the current internal ecological quality and current threat level of existing TSES elements. The proposals primarily focus on reconstruction of areas, which require further supplementation of vegetation, extension, change of species distribution, reconstruction of historical vegetation and historical parks, etc.

The output of the aforementioned proposals (a–c) is a proposal to create a functional skeleton of TSES. This is the central group of proposals and forms the core and substance of TSES: proposal for biocentres, biocorridors, interactive elements and buffer zones.

4.5.2 *Proposal for Ecostabilising Measures*

These proposals suggest implementation of the following ameliorative and technical measures:

(a) Proposal for localisation of new ecostabilising elements mainly focuses on the localisation of groups and strips of NFWV. This proposal group aims to enhance overall ecological stability. The proposal for ecostabilising measures needs to be formulated to improve the retention capacity of the landscape, hamper water run-off, increase the self-cleaning ability of the landscape and provide anti-erosion protection of landscape, as well as other perception-based values as aesthetic values (Štefunková and Cebecauer 2006). It draws on the map of abiocomplexes, which contains the criteria for the proposal for ecostabilising elements, e.g. slope dynamics zones (Špinerová 2010).

(b) Proposal for ecologically optimum utilisation of agricultural landscape—including agrotechnical, agrochemical and agromeliorative measures resulting from the analysis of abiocomplexes in the territory. These proposals partly overlap with the results of more complex, integrating management procedures, e.g. those of landscape-ecological planning LANDEP (Miklós 1990; Hrnčiarová and Miklós 1991; Špinerová 2010; Mederly et al. 2012). They can be further divided into the following measures:

- Proposal for creating optimum division of agricultural land (proposal for landscape-ecological optimum parcelling),
- Proposal for greening of the agricultural landscape and a proposal for crops necessary for planting vegetation, permanent grasslands on agricultural land,
- Proposal to create ecotone areas at the interface with ecologically significant landscape segments with the aim to protect these valuable territories from the negative impact of agricultural production,
- Proposal for grassing or afforesting the areas threatened and damaged by erosion,
- Proposal for measures in crop structure (focuses on selection of the most suitable structure of crops and exclusion of unsuitable crops, especially in terms of slope inclination, erodibility, water, air and salt regime of soil and topographic position),
- Proposal for agrotechnical measures (proposes the selection of the most suitable and exclusion of unsuitable agrotechnical measures, especially in terms of inclination, depth, skeletality, graininess of soils, etc.),
- Proposal for agrochemical measures (suggests the selection of most suitable or exclusion of unsuitable agrochemical measures including fertilisation, in terms of water and salt regime of soils, soil reaction and topographic position),
- Proposal for technical measures to mitigate erosion effects—proposal for safeguarding gully and furrow erosion, proposal to create anti-erosion terraces, retention ditches, implement drainage and foothill measures of slopes, and a proposal for soil decontamination.

In the agricultural landscape, the ecological stability of a territory depends primarily on the presented measures because the conditions for establishing biocentres and biocorridors are largely insufficient. Moreover, these ecostabilising measures ensure whole-territory character of TSES; however, even if a sufficiently dense network of biocentres and biocorridors is developed, the functionality of TSES may be paralysed by improper use of the surrounding area.

(c) Proposal for ecostabilising measures in the forest ecosystems derived from the synthetic map of abiocomplexes and map of the proposed TSES skeleton. The following principles should be applied in forest ecosystems:

- Proposal for safeguarding regeneration of natural species composition of stands,
- Differentiation of forest stands on their importance for TSES and their prospective forms of their use,
- Securing protection of endangered and autochthonous species in forest stands,
- Stabilisation of forest land borders,
- Proposal to prevent lowering of the rotation age of stands,
- Proposal for utilising logging and regeneration methods which would maximise reduction of the likelihood of damage to soil cover and herbaceous layer,
- Proposal for increasing diversity of forest ecosystems within transition zones using suitable indigenous woody species and gradual developing age and spatial structure of these stands by means of selection felling,
- Proposal for revitalisation of damaged forests,
- Proposal to change species composition of the stands with high resistance to air pollutants and introduce a spatial and temporal system for regeneration of the most pollutant-resistant forest elements (logging-regeneration procedures, modification of rotation age, etc.),
- Proposal for taking consistent and well-timed silvicultural measures in the establishment, tending and reproduction of stands,
- Proposal for surface stabilisation of stands in the establishment and tending stage with regard to adverse wind direction,
- Proposal to observe the principles of stand hygiene, especially in coniferous forests,
- Proposal for effective implementation of conservation measures—performing pest control and eliminating their negative impact,
- Proposal for differentiated tending with intensive measures, especially in young age of stand, with regard to the most pollutant-resistant elements of stands,
- Proposal for stabilisation of landslides in forest areas,
- Proposal for groundwork and retaining walls to avert erosion-accumulation processes.

(d) Proposal for hydroecological measures includes measures aimed at the protection of water sources and aquatic ecosystems. These measures are devised based on a map of abiocomplexes and a map of positive factors. The major hydroecological measures relevant for TSES include:

- Proposal to designate protection zones along aquatic ecosystems—mainly along natural watercourses to protect them from adverse environmental impact. This measure is particularly important in case of watercourses located in the agricultural and urbanised landscape. The literature on the subject asserts a required width of 10–15 m for such strip. The simplest solution for creating it suggests grassing it followed by natural succession,
- Proposal for revitalisation of watercourses—aimed at enhancing the ecological quality of disturbed and endangered aquatic ecosystems (regulated watercourses, inadequate structure of riparian vegetation, etc.). This group includes the following proposals: restoration and reopening of the remnants of an original watercourse bed, if preserved, reduction of the longitudinal gradient and slowdown of flow by increasing the length of a watercourse, restoration of the linkage between a watercourse and a floodplain (removal of impermeable reinforcement of river bed and banks), restoration of natural riparian vegetation, etc.,
- Proposal for grassing of significant water management areas—to provide protection of retention areas for drinking water sources, spring areas, water supply and important water management watercourses, etc. This proposal also suggests a change in the utilisation of protection zones of important water sources, in case the current landscape structure does not correspond with the protection of water resources, e.g. proposal to change the use of arable land by grassing or planting protective vegetation in sanitary protection areas around drinking water sources, etc.

(e) Proposal for ecostabilising measures in urban areas—aimed at improving the overall quality of the environment:

- Proposal for planting and reconstruction of residential vegetation (parks, housing estate, hygienic function, etc.),
- Proposal to eliminate stress factors of the urban environment as well as their negative impact,
- Proposal to attain favourable proportion of built-up areas and residential vegetation,
- Proposal for maintaining residential vegetation.

It should be noted that these are only framework measures and become effective in conjunction with other groups of proposals. They do not represent specific guidelines for soil cultivation, or a land consolidation project, but can constitute a basis for such types of projects.

4.5.3 Proposals for Legal Protection of the Elements of the Territorial System of Ecological Stability

The most natural tool providing legal protection to TSES elements is the enforcement of Act No. 543/2002 on nature and landscape protection, as amended, by declaring a TSES element a protected area of some category of territorial protection, namely a protected landscape area, national park and its protection zone, protected area and its protection zone, nature reserve and its protection zone, natural monument and its protection zone.

Proposal for legal protection of TSES elements can be formulated as follows:

(a) Inclusion of existing protected areas into the network of TSES and determining their functions—it is a confirmation of the current state of legal protection of individual TSES elements. The proposals were developed based on the confrontation of the maps of proposed skeleton of TSES and positive SEF. Each TSES element receives a certain level of protection (from 2. to 5. degree).
(b) Proposal to bolster legal protection—the proposal focuses on increasing of the level of protection of those TSES elements, whose current level of protection is not sufficient or fails to perform required functions of TSES elements. As well as in case of the former proposal group, the basis is for this one also arises from comparing the maps of the proposed TSES skeleton and positive SEF. Similarly, the level of protection should be stated here (from level 2 to 5).
(c) Proposal for designation of a new protected area addresses protection and the level of protection of the TSES element currently under no protection as well as protection of newly proposed elements of TSES. The proposal for legal protection is projected on the map of proposed TSES elements. Both the level of and the reason for protection need to be clearly specified.
(d) Proposal for protection of natural resources, especially forest and water resources—the proposal suggests recategorisation of commercial forests in the areas incorporated into TSES into categories of protective forests and special-purpose forests as stipulated by the Act No. 326/2005 Coll. on forests. The proposal to create a category of protective forests mainly draws on the interpretation of abiocomplexes, maps of area classification based on abiotic factors. This category should also encompass forests located in sensitive areas with the incidence of geodynamic processes. Special-purpose forests with their primary function of nature protection and mainly those forests act as interactive elements in the skeleton of TSES. The protection of TSES elements is significantly strengthened by the legal protection of water sources, namely large-scale protection of water sources in protected water management areas and catchment areas of water supply courses or sanitary protection zone of class II. as stipulated by the Act No. 364/2004. Also, other laws specifying protection of thermal and mineral water sources and soil sources positively affect TSES.
(e) Proposal for revoking protection—such cases are rare and are applied when a territory is under legal protection; however, currently the landscape-ecological value of the territory is low with high degree of degradation which prevents

it from performing required functions. The proposals are drafted based on the overlay of the map of area classification and synthesis map of positive factors.

The final output of these measures is a proposal for legal protection of particular TSES elements and natural resources.

4.5.4 Proposals to Eliminate Stress Factors

These measures are designed to mitigate the impact of stress factors on the landscape. The map of negative elements in the given territory constitutes the basis for the group of proposal. This group of proposals can be further subdivided into individual measures consisting of:

(a) Proposal for new technological measures aimed at reducing the production of contaminants, especially from the proposal for technological measures to reduce air pollutants, as well as the proposal for technological measures to reduce pollution of soil, water resources and eliminate sources of environmental noise load.
(b) Proposal to eliminate sources of stress factors concentrates on mitigating the impact or entire elimination of the primary sources of stress factors negatively affecting various landscape elements.
(c) Proposal for revitalisation of areas with environmental loads—the measures are intended to revitalise individual environmental elements and natural resources.
(d) Proposal to mitigate the barrier effect between anthropogenic landscape elements and the elements of TSES. This group of proposals is based on the map of the proposed skeleton of TSES and the map of negative factors in a given territory. The proposal addresses overcoming of these barriers; e.g. when ecological corridors are interrupted by roads which cannot be closed, it is necessary to project tunnels, underpasses or other types of ecoducts as required by the most sensitive species which cross the road. A negative impact of transport corridors is a significant problem for migrating fauna and is responsible for a number of roadkills and injuries. This step is considered the most important, monitored, visible with considerable influence on the general public.

4.5.5 Comprehensive Proposals for the Territorial System of Ecological Stability

TSES documentation comprises the following parts (Miklós et al. 2018):

(a) Textual part contains data on the submission of a project, delineation of a territory, description of analyses, syntheses, interpretations and classifications, evaluations, proposals and regulations for creating a functional TSES.

(b) Graphical part contains a set of maps with the purpose-arranged contents and combination spatial information. As TSES is a tool primarily addressing spatial relationships, maps are a critical output of TSES.

The TSES documentation is continuously supplemented, and therefore it is advisable to draft the TSES in the GIS environment.

An adequate arrangement and projection of the results into a map form are a crucial step not only in terms of their presentation, but also a subsequent implementation of the results. The results could be arranged into three maps:

(a) Map 1. The current landscape structure contains:

- REPGES,
- Physiognomic–ecological mapping units of CLS,
- Protected areas of nature and natural resources.

(b) Map 2. TSES proposals contain:

- The skeleton of TSES (biocentres and biocorridors) and their protection status,
- Proposal for localisation of non-forest woody vegetation,
- Proposal for ecostabilising measures.

(c) Map 3. Territorial system of stress factors (transparent overlay) contains:

- Sources of stress factors,
- Area affected by stress factors, cores and barriers.

There are numerous methods, including GIS, which can be applied to produce a graphical representation of the results.

A flow chart of the main elements of the TSES methodology is shown in Fig. 4.4.

4.6 Proposals for Passports

These proposals aim to produce model passport documents for all ecologically significant landscape segments and contain precise data structure necessary for preparing passportisation documents for TSES. The data structure indicates location of landscape elements, their area, ecological significance, reasons for protection, passportisation of the factors threatening those elements.

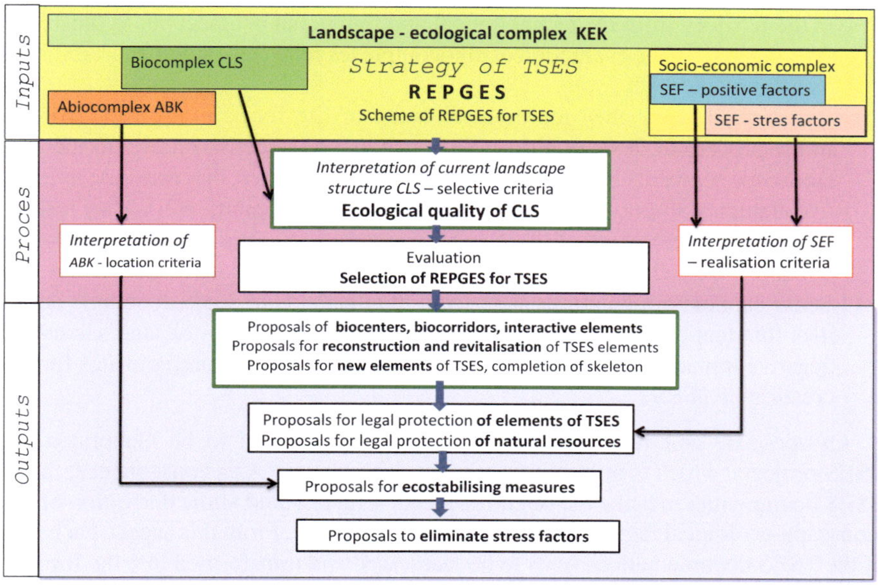

Fig. 4.4 Flow chart of development of TSES projects (Miklós et al. 2018)

4.7 Formation of Landscape-Ecological Regulations for Spatial Planning Documentations

TSES, as binding legislative material, is a part of the set of land-technical project documentation and must be transformed into the form of ecological regulation of the spatial optimisation of a territory. Pursuant to the law, a regulation of spatial arrangement and functional use of land is a binding directive, which guides the localisation and arrangement of a certain structure or implementation of a certain activity in a territory. It is expressed by the values of characteristics of elements of landscape structure in words, numbers and, if applicable, also graphically. The regulation has the character of bans, restrictions or supporting factors in relation to spatial arrangement and functional use of the territory. Hereby, the regulation determines forbidden, restricted and allowable activity or functions in a territory (Article 139a of Act No. Coll., 50/1976 as amended). The application of the regulations divides a territory according to the possibility of its use, conditional use or strict protection. On the one hand, the regulations either restrict or prohibit performance of certain activities, but on the other they allow some activities regarding utilisation of the territory.

Landscape-ecological regulations, with respect to the TSES, can be categorised into the following groups:

(a) Abiotic regulations: their values are conditioned by the properties of the abiotic complexes of the territory. They determine a localisation of potential activities in the territory from the technological and functional perspective. In the TSES documentation, the evaluation of abiocomplexes relies on a set of ecostabilising measures.

(b) Biotic regulations: they represent diversity of life forms in a given territory. In the process of designing the TSES, they act as selective criteria to identify landscape segments for the skeleton of TSES (biocentres, biocorridors, interactive elements, important gene pool sites, significant habitats, etc.). They restrict the spatial development of those socio-economic activities, which could pose a threat to TSES elements.

(c) Socio-economic regulations stem from the demands of socio-economic activities that hamper development of other activities by means of land seizure or negative impact. They have a character of implementation prerequisites for the creation of TSES.

An adequate and functional array of regulations needs to be elaborated for TSES projects, which is an exceptionally complex matter. As a consequence, much TSES documentation lacks the scientific rigour which would allow derivation of the landscape-ecological regulations for spatial development. From this aspect, each part of the TSES documentation needs to be reassessed and transformed into the form of a regulation, and they are subsequently reflected into individual project documents (land consolidation projects, spatial planning, river basin management plans, forest management plans, flood risk management plans). This implies that each area of a territory will have identified the activities which, from a TSES perspective, cannot be localised in a given area and can be localised with certain restrictions, as well as the activities which are appropriate and even desired.

Based on the aforementioned regulations, it is possible to identify a suitable activity in a given territory and steer the spatial development in such a way that it complies with the provisions of TSES.

The output of this phase is the spatial projection of elements of TSES as regulations in the form of limited, restricted and supporting activities.

References

Begon M, Harper JL, Townsend CR (1997) Ekologie: jedinci, populace a spoločenstva. Vyd. Univerzity Palackého, Olomouc, 949p

Bezák P, Izakovičová Z, Miklós L et al (2010) Reprezentatívne typy krajiny Slovenska. Bratislava, Ústav krajinnej ekológie SAV, 180p

Buček A, Lacina J, Löw J (1984) Teoretické východiská a typológia ÚSES. Pracovné materiály KRB, Agroprojekt Brno, p 12

Buček A, Lacina J (1984) Biogeografický přístup k vytváření územních systému ekologické stability krajiny. Zprávy Geografického ústavu ČSAV Brno 21(4):27–35

Diviaková A (2008) Mapovanie a hodnotenie líniových formácií nelesnej drevinovej vegetácie v katastrálnom území obce Žibritov. In: Špulerová J, Hrnčiarová T (eds) Ochrana a manažment poľnohospodárskej krajiny. Zborník príspevkov z konferencie, Ústav krajinnej ekológie SAV, Bratislava, pp 153–158

Ellenberg H (1974) Zeigewerte des Gefäßpflanzen Mitteleuropas. Scripta Geobotanica, 9, Göttingen, 97p

Feranec J, Oťaheľ J (2001) Krajinná pokrývka Slovenska/ Land cover of Slovakia. VEDA, Bratislava

Feráková V, Maglocký Š, Marhold (2001) Červený zoznam papraďorastov a semenných rastlín Slovenska. In: Baláž D, Marhold K, Urban P (eds) Červený zoznam rastlín a živočíchov Slovenska, Ochrana prírody 20:44–78

Halada Ľ, Topercer J, Mederly P, Kartusek V (1995) Systém ekologickej kvality krajiny – ďalší prístup k manažmentu krajiny. Životné prostredie, 29, 5:271-273

Hellmund PC, Smith D (2013) Designing greenways: sustainable landscapes for nature and people, Island Press, 288p

Hrnčiarová T (1998) Ekologická a kultúrna významnosť krajiny vyplývajúca z legislatívnych predpisov. In: Izakovičová Z, Kozová M, Pauditšová E (eds) Implementácia trvalo udržateľného rozvoja. Zborník príspevkov z konferencie, Ústav krajinnej ekológie SAV, Bratislava, pp 99–103

Hrnčiarová T (1999a) Krajinnoekologické plánovanie pomocou metodiky LANDEP a metodiky EÚK. Geografický časopis 51(4):399–413

Hrnčiarová T (1999b) Prepojenie metodiky LANDEP na metodiku EÚK. Bratislava, Životné prostredie 33(1):11–16

Hrnčiarová T, Izakovičová Z (2000) Environmental approaches to sustainable development. Združenie KRAJINA 21, Bratislava, 252p

Hrnčiarová T, Miklós L (1991) Morphometric indices in the interpretation of water and material motion dynamics illustrated on the example Dolná Malanta. Ecology (CSFR) 10(2):187–221

Hrnčiarová T, Miklós L et al (1991) Prírodno-socioekonomické regióny Slovenska. Krajinnoekologická pilotná štúdia. Ústav krajinnej ekológie SAV, Slovenská komisia ŽP, Bratislava, 49p

Izakovičová Z (1995) Ecological interpretations and evaluation of encounters of interests in landscape. Ekológia (Bratislava) 14:261–275

Izakovičová Z (1996a) Landscape territorial system. In: Management and protection of environment. Faculty of Natural Sciences Comenius University, Bratislava, pp 285–292

Izakovičová Z (1996b) Územný systém ekologickej stability – základ trvalo udržateľného rozvoja krajiny. Krajinné inžinierstvo, 2, ÚKIS, pp 4–14

Izakovičová Z (1997) Evaluation of the anthropogenic change of the landscape structure. Ekológia (Bratislava) 16:73–80

Izakovičová Z, Hrnčiarová T, Králik J, Liška M, Miklós L, Moyzeová M, Pauditšová E, Ružičková H, Šíbl J, Tremboš P (2000) Metodické pokyny na vypracovanie projektov regionálnych územných systémov ekologickej stability a miestnych územných systémov ekologickej stability. MŽP SR, Združenie KRAJINA 21, Bratislava, 155p

Izakovičová Z, Miklós L, Drdoš J (1997) Krajinnoekologické podmienky trvalo udržateľného rozvoja. Veda, Bratislava, p 183

Izakovičová Z, Miklós L, Moyzeová M et al (2011) Model reprezentatívnych geoekosystémov na regionálnej úrovni. Ústav krajinnej ekológie SAV, Bratislava, p 88

Jongman RHG (1996) Research priorities: scientific concepts and criteria. In: Perspectives on ecological networks. Eur Centre Nat Conserv Ser Man Nat 1:151–160

Jurko A (1990) Ekologické a socioekonomické hodnotenie vegetácie. Príroda, Bratislava, p 195

Kostrowicki AS (1982) Synanthropization as a result of environmental transformation. Memorabilia Zool 37:3–10

Kowarik I (1990) Some responses of flora and vegetation to urbanization in Central Europa. In: Sukopp H, Hejný S, Kowarik I (eds) Urban Ecology. SPB Acad. Publ, The Hague, pp 45–74

Krcho J (1965) Oslnenie reliéfu v ľubovoľnom uhle a čase a jeho znázornenie do máp pomocou izalumklín. Geografický časopis 17(1):19–40

Krcho J (1970) Zostrojenie máp časovej a uhlovej dynamiky oslnenia reliéfu graficko-numerickým spôsobom a pomocou samočinných počítačov. Geografický časopis 22(4):205–245

Krcho J (1973) Morphometric analysis of the relief on the basis of geometrical aspect of field theory. Acta Geogr Universitas Comenianae (Bratislava) 1:3–233

Kočická E (2000) Optimisation of the territory exploitation with an example of the complex physical geographic evaluation. Ekológia (Bratislava), 19: 92-98, Supplement 2/2000

Kočická E (2001) Geoekologická (komplexná fyzicko-geografická) mapa ako podklad pre hodnotenie ekologickej únosnosti krajiny. Acta Facultatis Ecologiae, Zvolen, 8:31-38

Kočická E (2007) Stanovenie ekologickej únosnosti na základe citlivosti krajiny na acidifikáciu (modelové územie povodie Belianskeho potoka – Štiavnické vrchy). Vydavateľstvo Technickej univerzity vo Zvolene

Kočický D et al (2008) Komplexný program protieróznej ochrany a návrh opatrení na zvýšenie retenčnej schopnosti územia SR v členení podľa čiastkových povodí. Záverečná správa. Esprit s.r.o, Banská Štiavnica

Kudrnovká O (1975) Morfometrické metódy a jejich aplikace při fyzickogeografické regionalizaci. Studia geographica 43, 182p

Lieskovský J (2006) Modelovanie horizontálnych vzťahov v hydrologických GIS analýzach. In: Faixová-Chalachanová J, Fencík R (eds) GIS vo vodnom hospodárstve, SAGI Bratislava (elektronický zborník)

Löw J et al (1984) Zásady pro vymezování a navrhování územních systémů ekologické stability v územně-plánovací praxi. Agroprojekt, Brno, p 55

Löw J et al (1995) Manual for designers of the local territorial system of ecological stability. Methods for elaboration of the documentation. Appendix, Brno (in Czech), 124p

Maglocký Š (1993) Hodnotenie biotopov. Životné prostredie, Vol. 27, No. 4:196-199

Marhold K, Hindák F (1998) Zoznam nižších a vyšších rastlín Slovenska. Veda Bratislava, 687p

Mederly P, Krautschneider J, Halada Ľ et al (2012) Projekt pozemkových úprav v k.ú, Jasenica. Všeobecné zásady funkčního usporiadadania územia. Nitra: Regioplán; Žilina: Ekoped

Míchal I et al (1991) Územní zabezpečování ekologické stability: Teorie a praxe. Praha, MŽP ČR, p 150

Michalko J et al (1986) Geobotanická mapa ČSSR. Slovenská republika. /Textová časť/, Veda, Bratislava, 168p + 40p prílohy

Miklós L (1986) Stabilita krajiny v ekologickom genereli SSR. Bratislava, Životné prostredie 20(2):87–93

Miklós L (1990) Ekologické plánovanie krajiny LANDEP I. Učebné texty, Slovenská technická knižnica pre MŽP SR a Krajinnoekologické centrum Banská Štiavnica, Bratislava, 75p

Miklós L (1991) Zásady slovenskej ekologickej politiky. Životné prostredie 25(4):174–178

Miklós L, Izakovičová Z (1997) Krajina ako geosystém. VEDA, SAV, Bratislava, p 152

Miklós L, Kozová M, Ružička M et al (1986) Ekologický plán využívania Východoslovenskej nížiny v mierke 1:25.000 (Ecological management plan of the Eastern lowlands at a scale of 1:25,000, in Slovakian). In: Ekologická optimalizácia využívania VSN, 3:5–312. ÚEBE CVEB SAV. Bratislava

Miklós L, Špinerová A, Belčáková I, Diviaková A (2018) Instruments and tools for sustainable land use and spatial development processes. In: Belčáková I, Miklós L (eds) Landscape considerations in spatial planning processes. Peter Lang: Frankfurt a.M. et al., ISBN 978-3-631-74971-5

Miklós M, Krcho J, Hrnčiarová T, Matečný I, Kozová M (1997) Morfometrické ukazovatele reliéfu a ich účelové interpretácie pre plánovacie procesy. Nadácia Katedry UNESCO pre ekologické vedomie, Banská Štiavnica, p 94

Miklós L (2002a) Share of ecological quality area per inhabitant in naturl settlement subregions. In: Landscape Atlas of Slovak Republic, Chapter VII, Map 6. MŽP SR Bratislava, SAŽP Banská Bystrica

Miklós L (2002b) Share of ecological quality area per inhabitant in naturl settlement microcroregions. In: Landscape Atlas of Slovak Republic, Chapter VII, Map 7. MŽP SR Bratislava, SAŽP Banská Bystrica

Miklós L, Izakovičová Z, Boltižiar M, Diviaková A, Grotkovská L, Hrnčiarová T, Imrichová Z, Izakovičová Z, Kočická E, Kočický D, Kenderessy P, Mojses M, Moyzeová M, Petrovič F, Špinerová A, Špulerová J, Štefunková D, Valkovcová Z, Zvara I, Hreško J, Lauko V, David S (2006) Atlas of representative geoecosystems of Slovakia. ÚKE SAV, Bratislava

Minár J (1992) The principles of the elementary geomorphological regionalization. Acta Facultatis Rerum Naturalium Universitas Comenianae, Geographica, Bratislava: PRIF UK, 33:185–198
Minár J (1995) Niektoré teoreticko-metodologické problémy geomorfológie vo väzbe na tvorbu komplexných geomorfologických máp. Acta Geographica Rerum Naturalium Universitatis Comenianae, Geographica 36:7–125
Minár J, Tremboš P (1994) Prírodné hazardy - hrozby, niektoré postupy ich hodnotenia. Acta Geographica Rerum Naturalium Universitatis Comenianae, Geographica 35:173–194
Minár J, Tremboš P (1995) Evaluation of the natural hazards in the landscape planning. Acta Environmentalica Universitatis Comenianae 4–5:211–222
Midriak R (1999) Ekologická únosnosť vysokohorských oblastí z hľadiska morfodynamických procesov. Bratislava, ÚKE SAV, Životné prostredie, 33, 1:42-45
Nowicki P, Bennet G, Middleton D, Rientjes S, Wolters R, (eds.) (1996) Perspectives on ecological networks. European Centre for Nature Conservation, series Man and Nature, vol. 1
Pasák V et al (1983) Ochrana zemědelské pôdy před erozí. Metodiky pro zavádění výsledku výskumu do zemědelské praxe. ÚVTIZ, 11, Praha, 77 pp
Pauditšová E, Reháčková T, Ružičková J (2007) Metodický návod na vypracovanie miestneho územného systému ekologickej stability. Acta Environmentalica Universitatis Comenianae 15(2):61–82
Polák P, Saxa A (eds) (2005) Priaznivý stav biotopov a druhov európskeho významu. ŠOP SR, Banská Bystrica, p 736
Raunkier C (1905) Types biologiques pour la geographie botanique. Bull Acad Sci Denmark Overs K Danske Vid Selsk Forhandl Kobenhavn
Ružička M, Miklós L (1982) Landscape ecological planning (LANDEP) in the process of territorial planning. Ekologia (CSSR) 1(3):297–312
Ružičková H, Ružička M (1973) Druhotná štruktúra krajiny ako kritérium biologickej rovnováhy. Quaestiones geobiologicae 12:23–62
Ružičková H, Halada Ľ, Jedlička L, Kalivodová E et al (1996) Biotopy Slovenska. Príručka k mapovaniu a katalóg biotopov. Ústav krajinnej ekológie SAV, Bratislava, 192p
Ružičková H, Topercer J, Halada Ľ, Šteffek J, Múdry P, Machanský S (1990) Interpretácia biotických zložiek pre krajinnoekologickú optimalizáciu územia Zamaguria – Ždiaru. Záverečná správa, Ústav krajinnej ekológie SAV, Bratislava, 117p
Ružičková H, Ružička M (1973) Druhotná štruktúra krajiny ako kritérium biologickej rovno¬váhy. Quaestiones geobiologicae 12:23–62
Smith D, Hellmund PC (eds) (1993) Ecology of greenways. Design and function of linear conservation areas. University of Minnesota Press, 222p
Špinerová A (2010) Krajinno-ekologické limity poľnohospodárskeho využitia povodia Ilijského potoka. Harmanec: VKÚ, a. s., 118p, ISBN 978-80-8042-613-2
Špinerová A (2015) Štruktúra krajiny ako regulátor dynamiky pohybu vody a materiálu. Zvolen, Vydavateľstvo TU vo Zvolene
Špulerová J (2004) Hodnotenie nelesnej drevinno-bylinnej vegetácie pre potreby krajinnoekologického plánovania. Dizertačná práca, ÚKE SAV, Bratislava, 107p
Stanová V, Valachovič M (eds) (2002) Katalóg biotopov slovenska. DAPHNE – Inštitút aplikovanej ekológie, Bratislava
Štefunková D, Cebecauer T (2006) Visibility analysis as a part of landscape visual quality assessment. In: Ekológia (Bratislava) : international journal of the biosphere, 25(1): 229-239
Tremboš P (1998a) Ekologická stabilita krajiny v katastrálnom území Dolná Trnávka. Geografické informácie 5:209–216
Tremboš P (1998b) Multikriteriálne hodnotenie ako metóda optimalizácie socioekonomických aktivít - niektoré metódy stanovenia váh kritérií. In: Izakovičová Z, Kozová M, Pauditšová E (eds) Implementácia trvalo udržateľného rozvoja. ÚKE SAV, Bratislava, pp 332–338
Tremboš P et al (1997) Miestne územné systémy ekologickej stability vo vybraných katastrálnych územiach v dosahu Závodu SNP, a.s. Žiar nad Hronom, 2. časť. Projekt MŽP SR, Bratislava, GEOINFO, Piešťany

Turner MG, Gardner RH (eds) (1991) Quantitative methods in landscape ecology: the analysis and interpretation of landscape heterogeneity, Springer 35(4):231–240

Tremboš P (1994) Identifikácia, charakteristika a interpretácia abiokomplexov v procese tvorby dokumentov regionálnych územných systémov ekologickej stability. Acta Facultatis Rerum Naturalium Universitatis Comenianae, Geographica, 35:157–171

Troumbis AZ, Farjon JMJ, Felton M, Miklós L, Mugica M, Jongman R (1995) Buffer zones and ecological corridors. A report to the subproject MN 2.7. European Topic Centre "Nature", European Centre for Nature Conservation,Tilburg

Wascher D, Jongman R (2000) European landscapes, classification, assessment and conservation. European Environmental Agency, Coppenhagen

Chapter 5
Conclusion

Current policies for the protection of nature and natural resources promote complexity and integrated approaches. Since their inception, territorial systems of ecological stability were considered a part of the comprehensive concept of landscape-ecological planning (LANDEP) (Ružička and Miklós 1982; Miklós 1996), as well as an essential part of all spatial planning processes. Therefore, they also satisfy the notions of integrated landscape management stipulated by original reading of the Act No. 7/2010 on flood protection in § 9, paragraph (1):

> In order to achieve optimal protection against floods, a proposal for measures under § 8 paragraph 4, flood risk management plan and the plan for river basin management must be coordinated with other instruments of spatial planning, particularly with land consolidation projects, land-use plans and forest management plans to constitute a tool of integrated landscape management throughout the entire area of the river basin.

The TSES is a key component of integrated land management, and it can be perceived as the most successful landscape-ecological concept incorporated into the ecological policy, legislation and practice in the SR after 1989.

The real importance of the TSES is secured by implementation in the following acts:

(a) Act on Nature and Landscape Protection 284/1994 and 543/2002 Coll.: there is the basic definition of TSES and its determination as a basic document for different planning.
(b) Act on Territorial Planning and Construction Order, amendments 262/1992 Coll. and 237/2000 Coll.: defines that the elements of TSES are obligatory regulation at all levels of territorial plans.
(c) Act on Land Arrangement and Land Ownership 331/1991 Coll. and its amendment 549/2004 Coll.: defines that TSES is an obligatory base and part of each land arrangement project; moreover, the need for improving the TSES function might be accepted as a legal cause for enactment of the land arrangement procedure.

L. Miklós et al., *Ecological Networks and Territorial Systems of Ecological Stability*, https://doi.org/10.1007/978-3-319-94018-2_5

(d) Act on Environmental Impact Assessment 127/1994 Coll. and 24/2006 Coll.: defines that TSES is an obligatory object of impact assessment.
(e) The Water Act 364/2004 Coll. (based on Water Framework Directive 2000/60 of EP and EC): forces the utilisation of the water protecting function of TSES with the coordination of water management tasks.
(f) Act 7/2010 Coll. on Flood Protection: enacts that the long-term management plan of watersheds should project also TSES, important landscape elements and the ecostabilising measures.
(g) TSES projecting became the subject of authorisation under Act 138/1992 Coll. on Authorised Architects and its later amendments.

In view of the above, the education of specialists for compiling TSES projects deserves special attention. This, however, also requires an adequate body of scientific and professional literature, which would adequately illuminate the theoretical and methodological aspects of the importance of ecological networks in the landscape, methodological procedures applied in the development of such networks, as well as the framework and possibilities of the implementation of the projects of ecological networks into actual spatial planning processes.

The content of the presented study aims to satisfy these requirements. It is up to the reader to decide whether its composition, style and language facilitate the spread of knowledge on ecological networks and especially the concept of territorial systems of ecological stability.

References

Miklós L (1996) The concept of the territorial system of ecological stability in Slovakia. In: Jongmann RHG (ed) Ecological and landscape consequences of land-use change in Europe. ECNC publication series on Man and Nature 2, Tilburg
Ružička M, Miklós L (1982) Landscape ecological planning (LANDEP) in the process of territorial planning. Ekologia (CSSR) 1(3):297–312

Appendices

The appendices are included to the book in order to illustrate the key outputs of TSES methodics on concrete model territories—the spatial distribution of the concerned issues on the maps. The outputs are on two most used hierarchical level, namely on the level of

- Regional TSES for the district Trnava (capital city of the county Trnava West Slovakia)—Appendix A
- Local TSES for the community Hody (district Galanta, West Slovakia) Appendix B.

L. Miklós et al., *Ecological Networks and Territorial Systems of Ecological Stability*, https://doi.org/10.1007/978-3-319-94018-2

Appendix A

The document of the regional territorial system of ecological stability RTSES is focused on the protection of diversity of conditions and life forms and on achieving ecological stability usually at the level of the district. It serves as an obligatory regulative for the

- Territorial plan of the region,
- Preparing documents for local TSES for the communities in the region
- Decision making of nature protection bodies and for
- Practical management of specially protected parts of nature and landscape (MoE Decree No. 24/2003, § 22, Section 3–6)

The document of the regional territorial system of ecological stability contains:

- Territorial projection of nature and landscape protection with an overview of selected parts of nature and landscape, which are of particular importance for
- Preservation of biological diversity and ecological stability of the landscape regardless of their protection level (hereinafter as "specially significant parts of nature and landscape"),
- Territorial projection of load on nature and landscape with assessment of a load degree and changes of landscape triggered by human activity,
- Territorial projection of ecological stability of landscape with categorisation by the proportion of specially significant parts of nature and landscape and degree of their anthropogenic changes,
- Proposal of the measures specifying regulations for nature and landscape protection and development.

The document of the regional territorial system of ecological stability is elaborated usually for a territory of a district, apart from the territory of a protected landscape area or a national park, including its protection zone, where the function of the regional territorial system of ecological stability is fulfilled by documents as specified by § 21 Section (1) of MoE Decree No. 24/2003. Thus, the RTSES

L. Miklós et al., *Ecological Networks and Territorial Systems of Ecological Stability*, https://doi.org/10.1007/978-3-319-94018-2

constitutes the fundamental landscape-ecological basis for the elaboration of the land-use planning documentation of a higher regional unit, and for other departments and their relevant documentations (nature conservation, territorial development, land consolidation projects, forest management plans river basins management plans a.o).

The map scale of the document of the RTSES usually corresponds with the map scale of the territorial plans; most often are represented on maps on a scale 1:50 000.

Examples of map outputs of the project of regional territorial system of ecological stability of the Trnava district (Fig. A.1, A.2, A.3, A.4, A.5, A.6, A.7, A.8 and A.9).

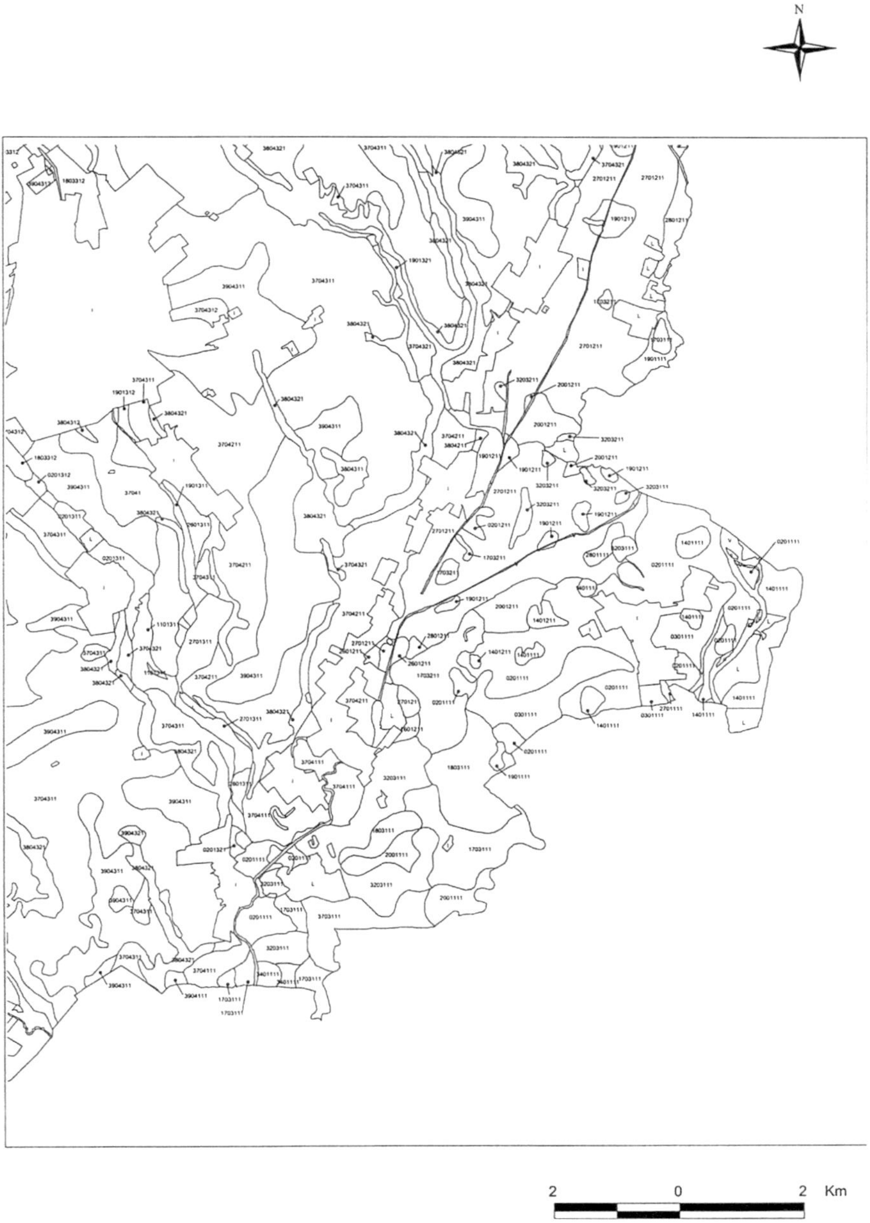
N
2
0
2 Km

◀ **Fig. A.1** Section of the map of abiocomplexes (Izakovičová et al. 2001).Each area of abiotic complex is characterized by 7 digits code $x_1x_2x_3x_4x_5x_6x_7$ of the values of five properties as:

$x_1x_2$00000 –subcomplex soil
00$x_2x_3$000 –geological base (bedrocks)–soilforming substratum complex
0000$x_5$00 –complex relief
00000$x_6$0 –slope angle
000000x_7 –subcomplex climate

As for example, the code of the abiocomplex 1401211 in an area means, that the abiocomplex has the following values of properties:

14 $x_3x_4x_5x_6x_7$ –abiocomplex with mollic fluvisol (typical), shallow, on fluvial sediments, moderately-heavy to light soil
x_1x_2**01**$x_4x_5x_6x_7$ –abiocomplex with loamy to sandy-loamy floodplain sediments
$x_1x_2x_3x_4$**2**x_6x_7 –abiocomplex on alluvial plain
$x_1x_2x_3x_4x_5$**1**x_7 –abiocomplex with slope angle 0 to 1°
$x_1x_2x_3x_4x_5x_6$1 –abiocomplex in warm, dry, moderately cold climatic region, with longer sun radiation

Since the figure is just an illustration of an example of a concrete model territory in Slovakia, the explanation of the full set of values is not essential. It is given in the work Izakovičová et al. 2001

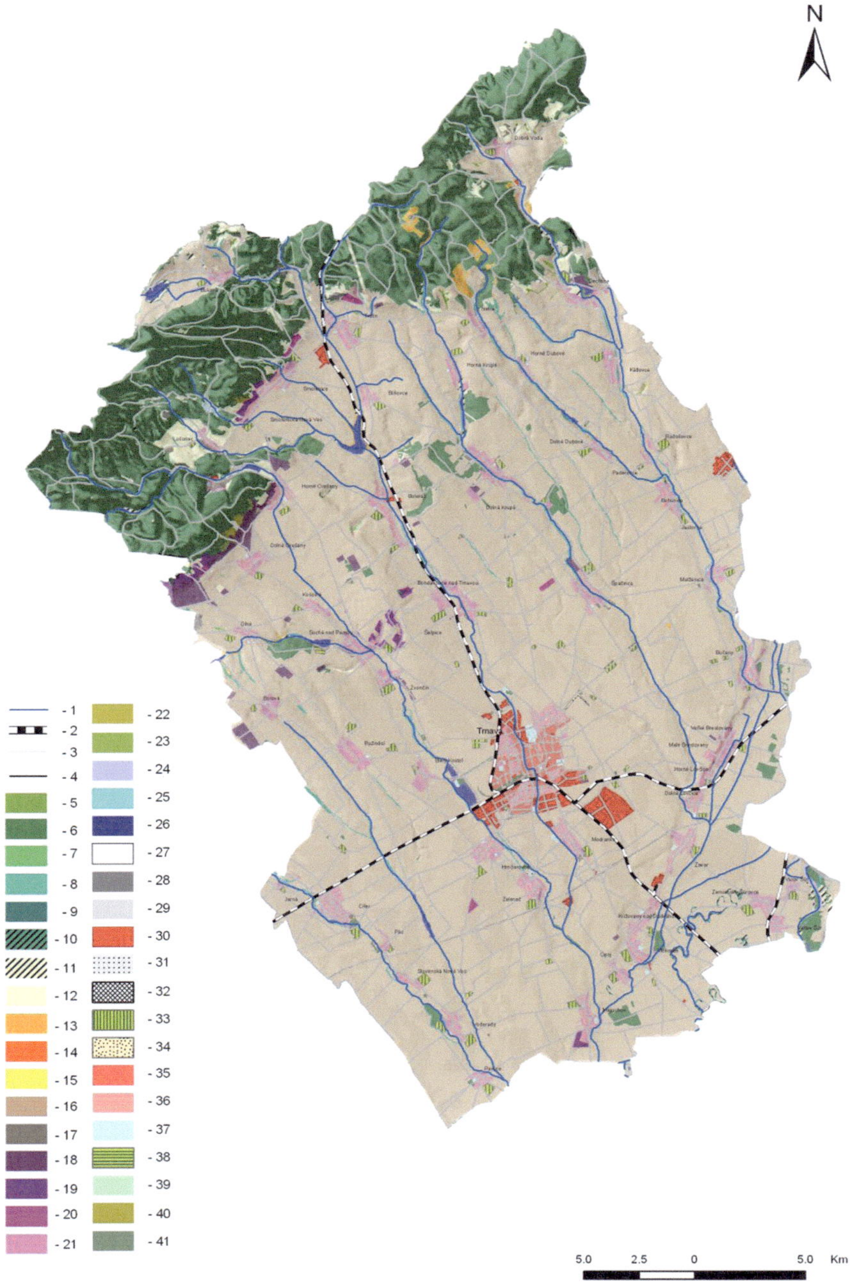
N
Trnava
- 1
- 2
- 3
- 4
- 5
- 6
- 7
- 8
- 9
- 10
- 11
- 12
- 13
- 14
- 15
- 16
- 17
- 18
- 19
- 20
- 21
- 22
- 23
- 24
- 25
- 26
- 27
- 28
- 29
- 30
- 31
- 32
- 33
- 34
- 35
- 36
- 37
- 38
- 39
- 40
- 41
5.0 2.5 0 5.0 Km

◀ **Fig. A.2** Secondary landscape structure/land cover.

1–water flows, **2**–railways, **3**–roads, **4**–motorways, **5**–forests, **6**–non forest natural wood vegetation, **7**–non forest artificial wood vegetation, **8**–bank vegetation, **9**–line vegetation artificial, **10**–wetlands, **11**–damp meadows, **12**–extensive grasslands, **13**–intensive meadows, **14**–intensive pastures, **15**–dry grasslands, **16**–arable land with big blocks, **17**–arable lands with small blocks, **18**–big-blocks vineyards, **19**–small-blocks vineyards, **20**–orchards, **21**–gardens, **22**–mosaic of the arable lands and grasslands, **23**–mosaic of the grasslands and non forest vegetation, **24**–water flows natural, **25**–water flows regulated, **26**–water reservoirs, **27**–natural rocks, **28**–uncover areas, **29**–other areas without vegetation, **30**–industry areas, **31**–mining areas, **32**–dump of industry waste, **33**–agricultural farms, **34**–field airport, **35**–urban areas, **36**–village areas, **37**–recreational areas, **38**–gardens with cottages, **39**–cottages areas, **40**–abandoned areas, **41**–transport areas

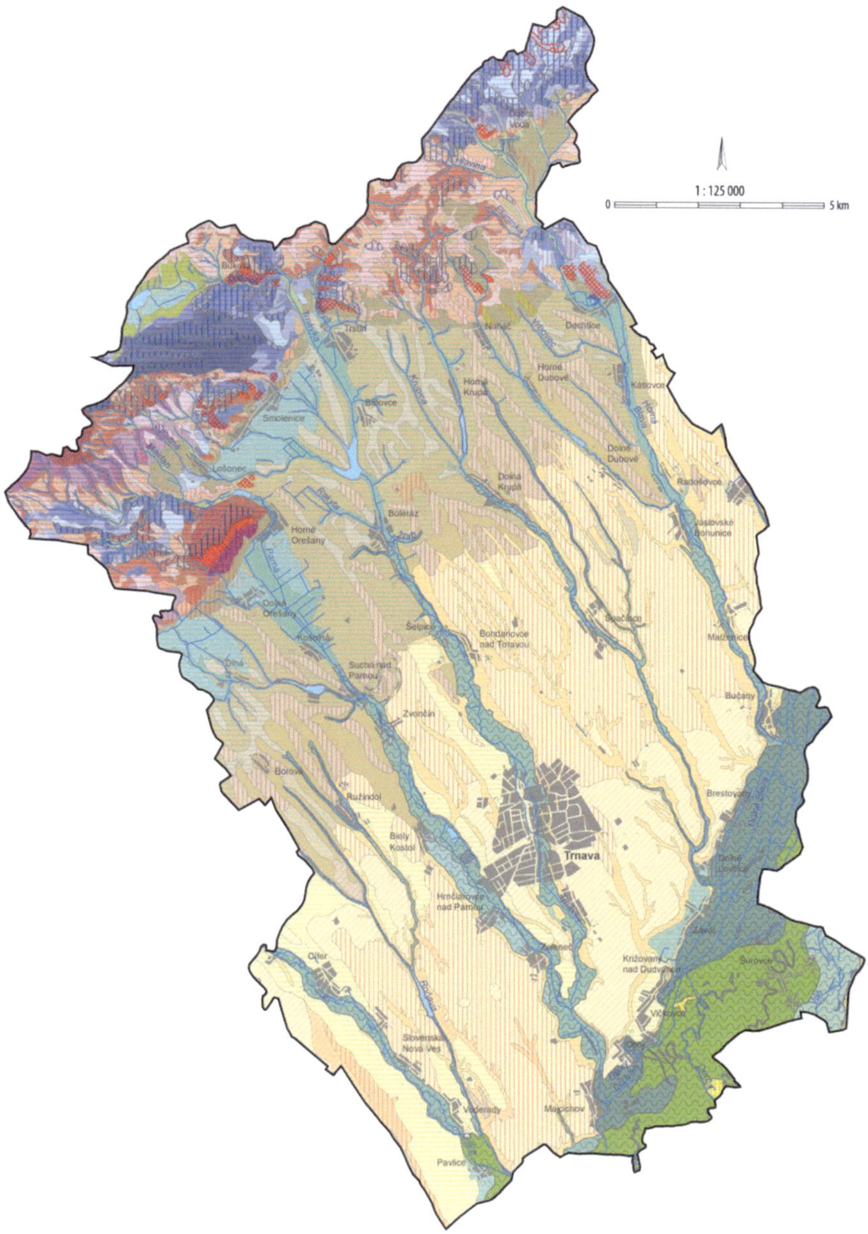

Fig. A.3 Types of representative geoecosystems. The legend in graphic form is attached as part of the Fig. A.3

Legend - representative ecosystems

	Abiotické podmienky	Azonálne spoločenstvá				Bioklimatcké podmienky charakterizované zonálnymi spoločenstvami												plocha [km²]
		Lužné lesy				Dubovo-hrabové lesy		Dubové lesy				Bukové lesy						
		Vŕbovo-topoľové nížinné lužné lesy	Dubovo-brestovo-jaseňové nížinné lužné lesy	Podhorské a horské jelšové lužné lesy	Slatiny	Dubovo-hrabové lesy karpatské	Dubovo-hrabové lesy panónske	Teplomilné submediterán-ne dubové lesy a skalné stepi	Teplomilné pontícko-panónske dubové lesy	Dubovo-cerové lesy	Dubové nátržníkové lesy	Kyslomilné bukové lesy	Vápnomilné bukové lesy	Bukové a jedľovo-bukové kvetnaté lesy	Bukové kvetnaté lesy podhorské	Kyslomilné bukové lesy podhorské		
Roviny	rovinná depresia		100.00%															27
Roviny	nečlenená rovina aluviálnych nív	100.00%																5
Roviny	horizontálne rozčlenená rovina agradačných valov		100.00%															27
Roviny	zvyšky meandrov a zamokrených depresií				100.00%													3
Roviny	nivné roviny na dnách nížinných pahorkatín		60.88%	39.12%														63
Roviny	nivné roviny na dnách vrchov a hornatinných dolín			100.00%														11
Roviny	slabo až mierne zvlnená rovina riečnych terás a náplavových kužeľov		13.57%			70.45%	15.13%			0.85%								23
Roviny	slabo zvlnená rovina pieskových pokrovov									100.00%								1
Roviny	slabo zvlnená rovina sprašových tabúľ						18.63%		24.32%	57.05%								218
Pahorkatiny	slabo členitá pahorkatina úvalín a svahov sprašových tabúľ						83.16%		10.19%	6.65%								66
Pahorkatiny	stredne členitá sprašová pahorkatina					76.16%	22.73%			0.88%	0.22%							96
Pahorkatiny	ploché chrbty sprašových pahorkatín					28.67%	6.21%		2.52%	62.60%								52
Pahorkatiny	stredne členitá polygénna pahorkatina (svahy terás, kužeľov, polygénna pahorkatina)					76.13%				21.38%				2.49%				3
Pahorkatiny	silne členité pahorkatiny a plošinaté predhoria					82.44%				10.04%	1.03%		2.84%	3.64%				17
Vrchoviny	mierne členené krasové vrchovinové plošiny a planiny					20.11%				13.24%	7.35%		22.46%	33.81%	3.03%			8
Vrchoviny	mierne členené vrchovinové plošiny a planiny na kryštalických horninách					60.83%				11.12%				28.04%				2
Vrchoviny	mierne členené vrchovinové plošiny a planiny na pestrých horninách					24.00%		1.11%		10.17%	0.54%		9.79%	54.39%				14
Vrchoviny	stredne členitá krasová vrchovina (ploché chrbty a mierne svahy)					31.51%				7.94%	5.36%		27.02%	23.83%	4.34%			14
Vrchoviny	stredne členitá vrchovina (ploché chrbty a mierne svahy) na kryštalických horninách					40.73%				12.94%				45.74%		0.59%		3
Vrchoviny	stredne členitá vrchovina (ploché chrbty a mierne svahy) na pestrých horninách					37.58%		0.85%		4.79%	0.58%	0.64%	9.87%	45.19%		0.51%		28
Vrchoviny	silne členitá krasová vrchovina (svahy polygénnych vrchovín)					14.64%				3.77%	8.78%		46.27%	24.57%	0.99%	0.98%		21
Vrchoviny	silne členitá vrchovina (svahy polygénnych vrchovín) na kryštalických horninách					11.35%		9.50%		1.52%				65.87%		11.77%		4
Vrchoviny	silne členitá vrchovina (svahy polygénnych vrchovín) na pestrých horninách					21.98%		1.91%		2.74%	3.98%	0.63%	19.52%	48.68%		0.56%		20
Hornatiny	stredne členitá krasová nižšia hornatina (masívne chrbty a svahy)												45.36%	54.64%				10
Hornatiny	stredne členitá nižšia hornatina (masívne chrbty a svahy) na kryštalických horninách									100.00%								1
Hornatiny	stredne členitá nižšia hornatina (masívne chrbty a svahy) na pestrých horninách					14.30%					18.65%	0.77%	27.71%	19.29%		19.28%		6
	Spolu %	0.63%	12.90%	4.80%	0.38%	20.49%	16.71%	0.15%	8.22%	23.35%	0.76%	0.05%	4.02%	7.14%	0.14%	0.27%		743

Fig. A.3 (continued)

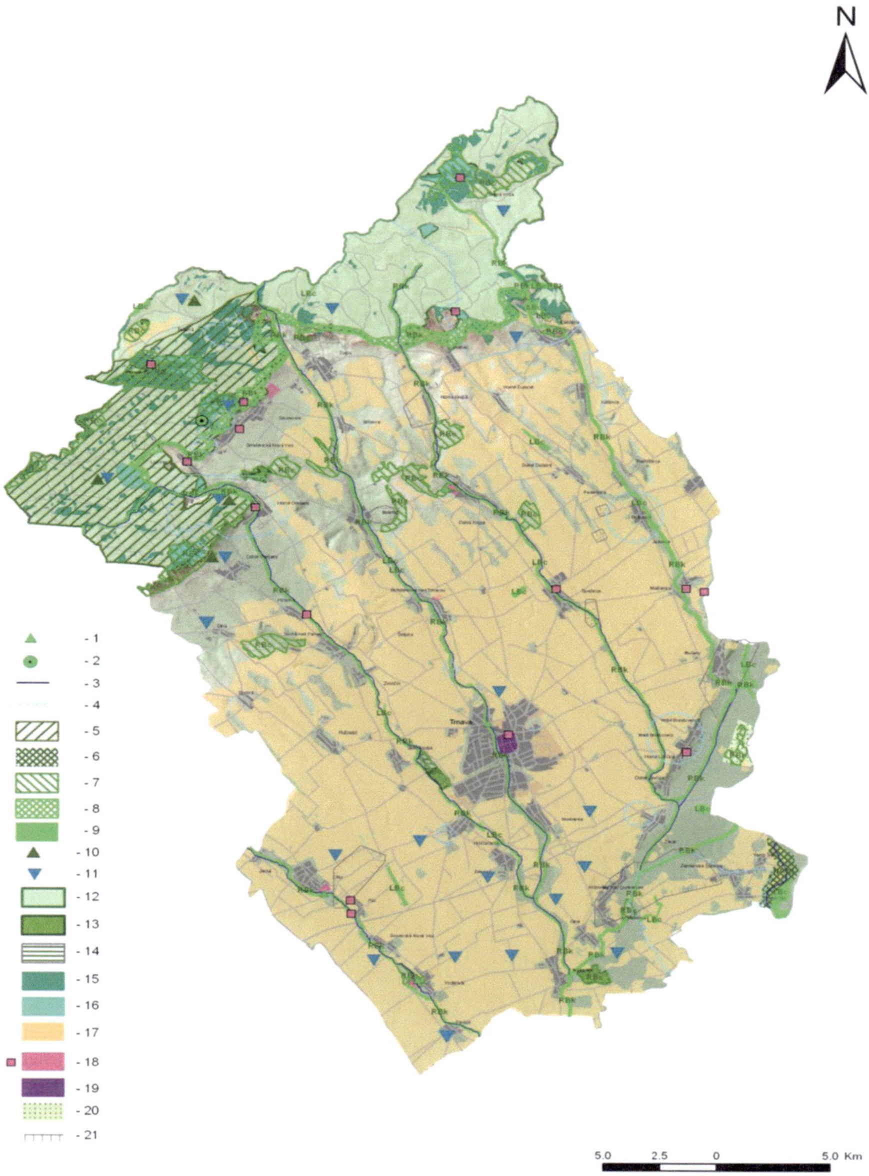

Fig. A.4 Positive socio-economic factors.

1–protected trees, **2**–carst, **3**–protected water flows, **4**–protected zones of the water resources, **5**–supraregional biocentre, **6**–supraregional biocorridor, **7**–regional biocentre, **8**–regional biocorridor, **9**–local biocentre, **10**–area of the NATURA, **11**–protected bird area, **12**–protected landscape area, **13**–protected area, **14**–puffer zone of the protected area, **15**–protected forests, **16**–forests with special purpose, **17**–protected soils, **18**–cultural monuments, **19**–town reservation, **20**–pheasant area **21**–protected area of the mineral resources

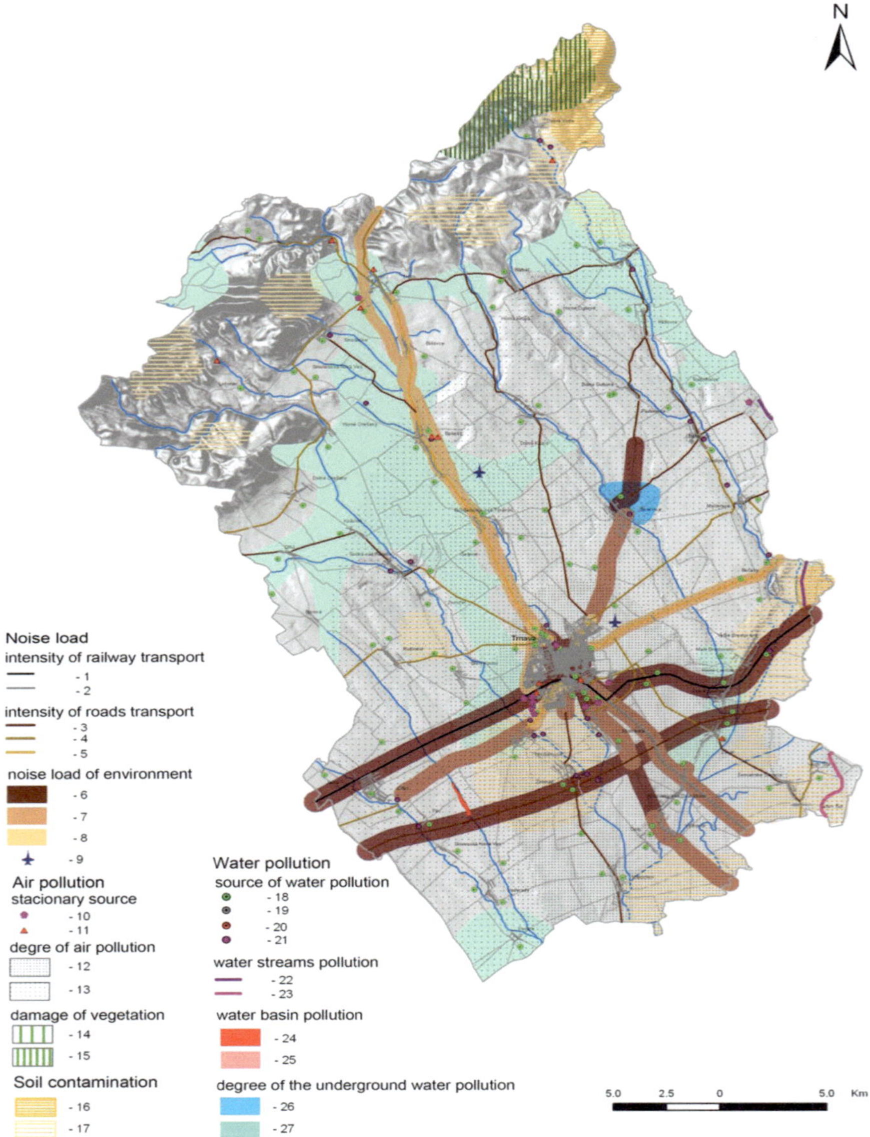

Fig. A.5 Anthropogenic stress factors.

Noise load: **1**–high, **2**–medium, **3**–very high, **4**–high, **5**–medium, **6**–very high, **7**–high, **8**–medium, **9** – airport
Air pollution: **10**–big, **11**–other important, **12**–high, **13**–medium, **14**–high, **15**–low
Soil contamination: **16**–very high, **17**–high
Water pollution: **18**–from agriculture, **19**–from industry and transport, **20**–dump of the waste, Water pollution: **18**–from agriculture, **19**–from industry and transport, **20**–dump of the waste, **21**–direct place of the drop of the polluted water, **22**–I. degree of pollution, **23**–III. degree of pollution, **24**–III. degree of pollution, **25**–II. degree of pollution, **26**–high, **27**–medium–direct place of the drop of the polluted water, **22**–I. degree of pollution, **23**–III. degree of pollution, **24**–III. degree of pollution, **25**–II. degree of pollution, **26**–high, **27**–medium

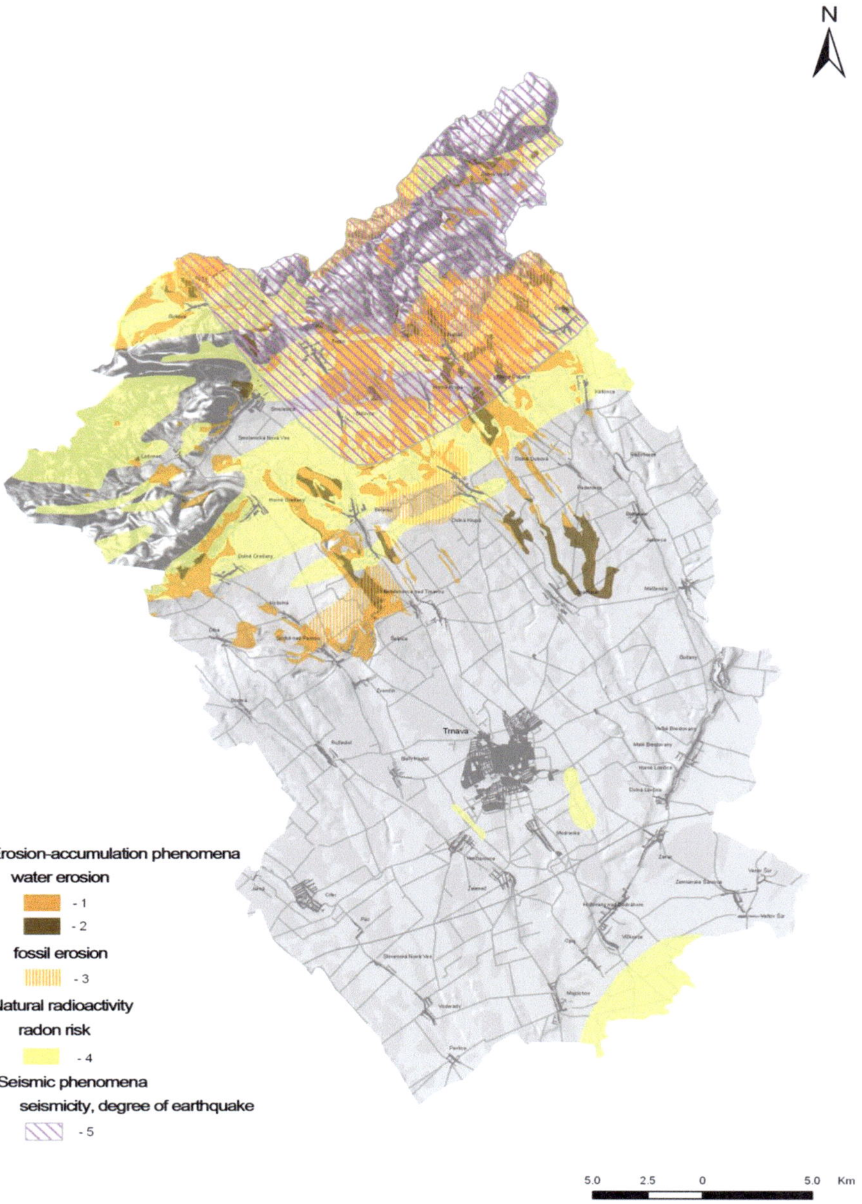

Fig. A.6 Natural stress factors.

Erosion-accumulation phenomena: **1**–medium water erosion, **2** – strong water erosion, **3**–fossil erosion
Natural radioactivity: **4**–radon risk
Seismic phenomena: **5**–potential degree of earthquake

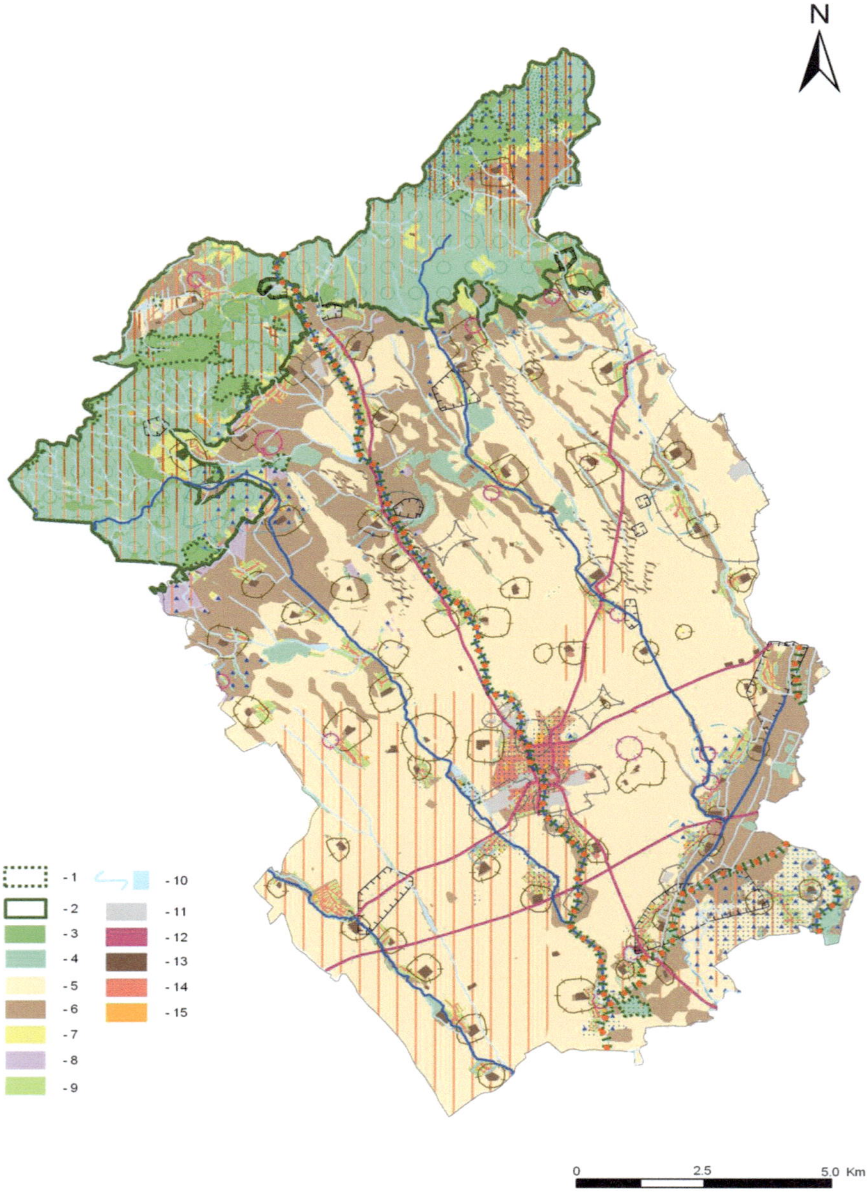

Fig. A.7 Conflict of interests in the landscape—environmental problems.
The input characteristics

1–Small scale protected area, **2**–Protected landscape area Malé Karpaty, **3**–Special-purpose forests, **4**–Productive and other forests and non-forest woody vegetation, **5**–Protected soil, **6**–Other types of soil, **7**–anent grassy vegetation, **8**–Vineyards, **9**–Gardens and orchards, **10**–Watercourses and all surface water, **11**–Industrial areas, **12**–Waste dump, **13**–Agricultural areas, **14**–Residential areas, **15**–Recreational areas

Matrix of conflict of interests:
The rows present the environmentally positive factors
The columns present environmentally negative factors
The squares present the real existing encounters of positive and negative factors in the model territory.

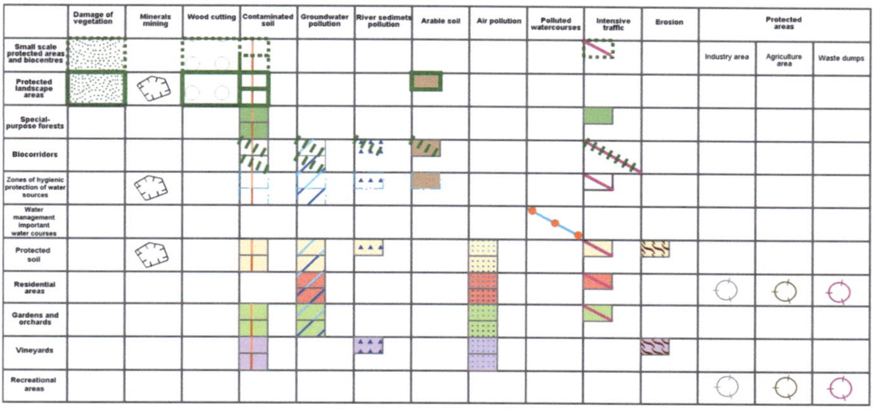

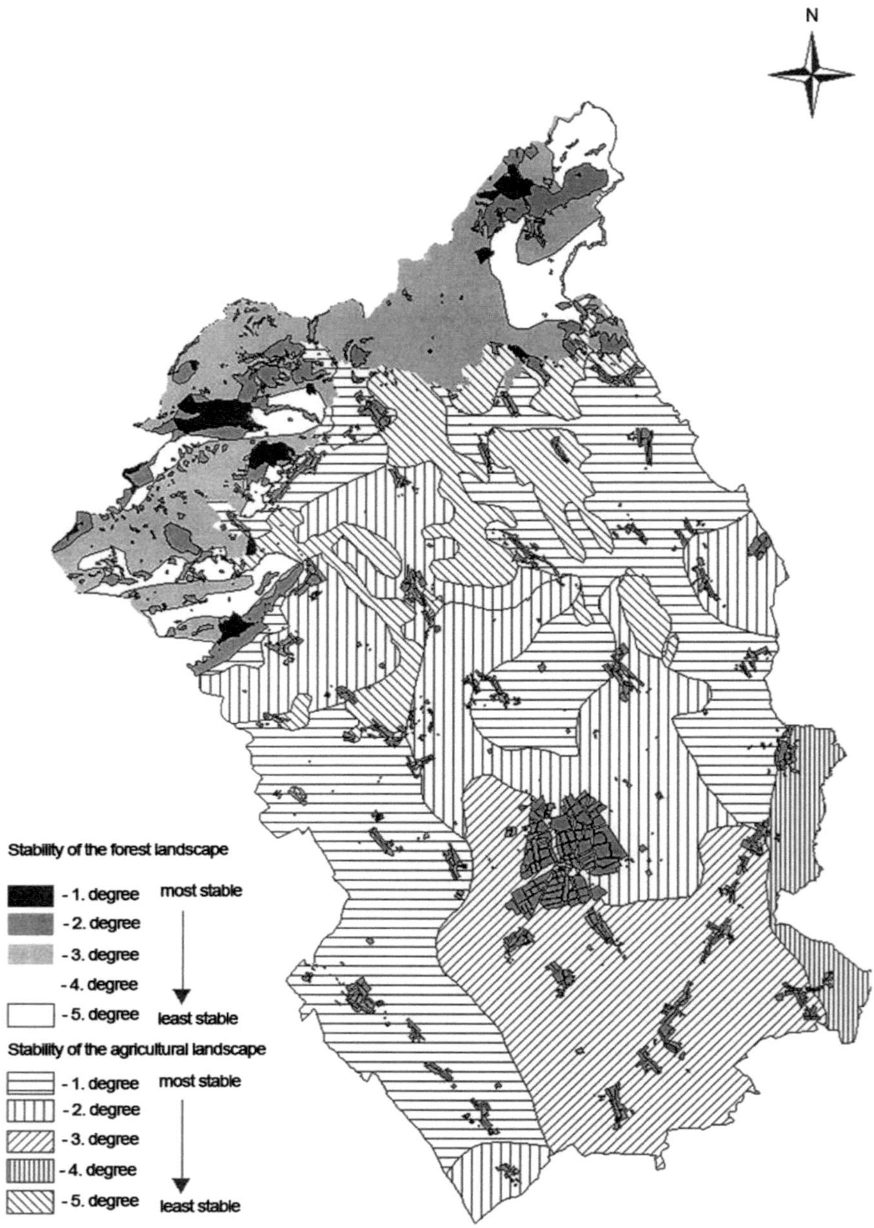

Fig. A.8 Ecological stability of the landscape in the model territory.
Explanation included also in the figure, as:

Stability of the forest landscape in 5 degree: **1**–most stable, **5**–least stableStability of the agricultural landscape in 5 degree: **1**–most stable, **5**–least stable

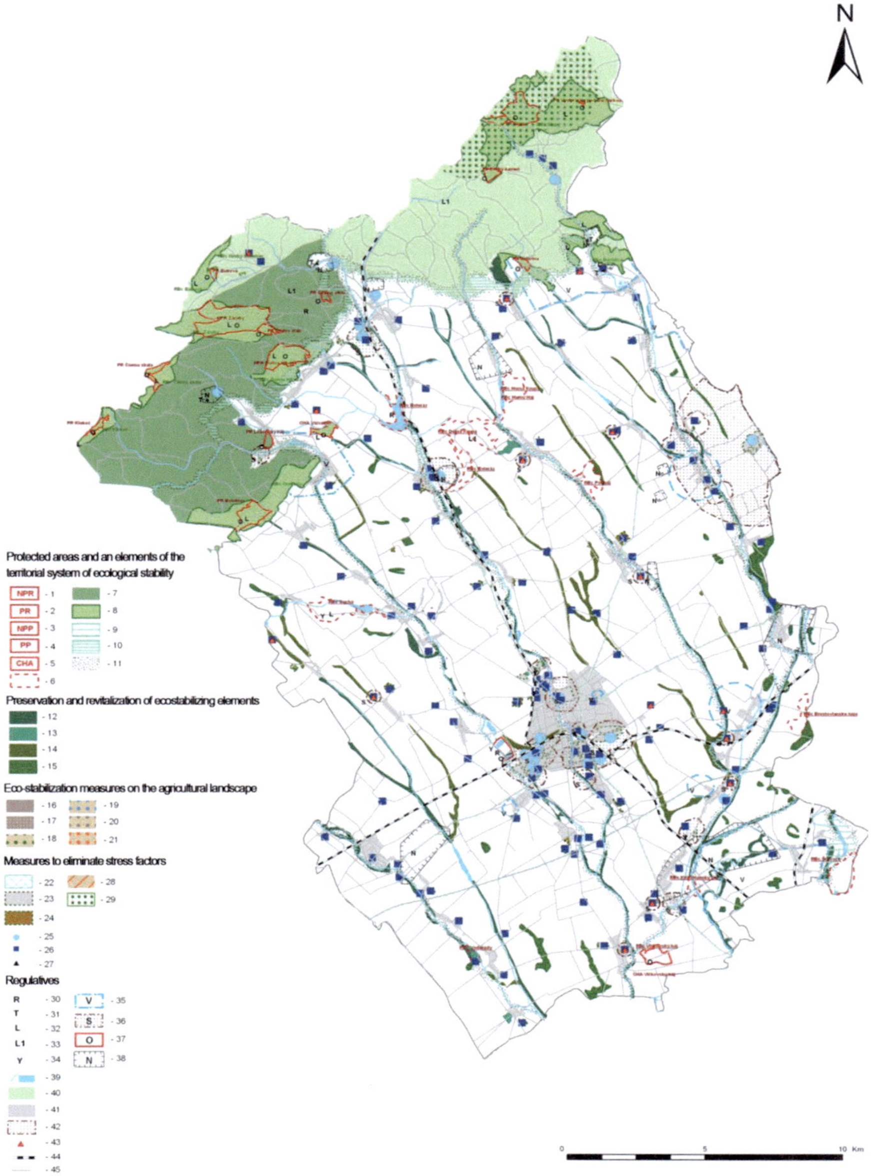
N
Protected areas and an elements of the territorial system of ecological stability
NPR - 1
PR - 2
NPP - 3
PP - 4
CHA - 5
- 6
- 7
- 8
- 9
- 10
- 11
Preservation and revitalization of ecostabilizing elements
- 12
- 13
- 14
- 15
Eco-stabilization measures on the agricultural landscape
- 16
- 17
- 18
- 19
- 20
- 21
Measures to eliminate stress factors
- 22
- 23
- 24
- 28
- 29
- 25
- 26
- 27
Regulatives
R - 30
T - 31
L - 32
L1 - 33
Y - 34
V - 35
S - 36
O - 37
N - 38
- 39
- 40
- 41
- 42
- 43
- 44
- 45
0
5
10 Km

◀ **Fig. A.9** Proposals and regulations.

1–National nature reserve–5. level of protection, **2**–Natural reserve–5. level of protection, **3**–National nature monument–5. level of protection, **4**–Nature monument–5. level of protection, **5**–Protected site–4. level of protection, **6**–Suggested protected sites, **7**–Over-regional biocentre, **8**–regional biocentre, **9**–Over-regional biocorridor, **10**–Regional biocorridor, **11**–Buffer zone in the vicinity of biocorridor, **12**–Riparian vegetation, **13**–Park vegetation, **14**–Line eco-stabilization and protective vegetation, **15**–Small groups of trees, **16**–Small scale arable soil, **17**–Small scale arable soil with anti-erosion measures, **18**–Special cultivation of Agricultural soil fund (soils) according to protection of nature, **19**–Special cultivation of ASF according to protection of water sources, **20**–Special cultivation of soils according to protection of water sources and deposit area, **21**–Special cultivation of soils in highly encumbered areas (pollution of water, soil and air), **22**–Plantation of hygienic-insulating vegetation in the vicinity of traffic, **23**–Plantation of hygienic-insulating vegetation in the vicinity of industrial areas, **24**–Plantation of hygienic-insulating vegetation in the vicinity of sources of bacterial contamination, **25**–Realize technical measures for elimination of air pollution sources, **26**–Realize technical measures for elimination of water pollution sources, **27**–Consider further mining possibilities of minerals, **28**–Realize research of real soil contamination and ensure its decontamination, **29**–Realize research of real vegetation damage and ensure its revitalization, **30**–Harmonize recreation with nature protection, **31**–Minerals mining with protection of nature, **32**–Exclude industrial wood cutting, **33**–Find balance between wood cutting and protection of nature, ensure re-categorisation of forests to the category of special purpose forests gradually, **34**–Harmonize using of landscape with protection of water sources, **35**–Harmonize using of protected range with hygienic requirements, **36**–Ensure measures following from Nature protection act, **37**–Find balance between using of landscape and deposit territory protection, **38**–Find balance between fishing and nature protection, **39**–Watercourses and all surface water, **40**–Forest vegetation, **41**–Residential part, **42**–Protected ranges of agricultural and technical objects, **43**–Waste dumps, **44**–Railways, **45**–Roads

Reference

Izakovičová Z et al (2001) Regionálny územný systém ekologickej stability okresu Trnava, Ústav krajinnej ekológie SAV, 127p

Appendix B

The document of the **local territorial system of ecological stability MUSES** is a basic document of protection of diversity of conditions and life forms, and for achieving ecological stability at the local level.

It forms the obligatory basic material for the

- Municipal territorial plan,
- Territorial plan of municipal zones,
- Land consolidation projects (MoE Decree No. 24/2003, § 22, Section 7–8).

The map scale of the document of the RTSES usually corresponds with the map scale of the territorial plans; most often are represented on maps on a scale 1:10,000 or 1:5000.

Example of map outputs of the project of the local territorial system of ecological stability of the cadastre area Hody (Diviaková et al. 2005) (Fig. B.1, B.2, B.3, B.4, B.5, B.6, B.7 and B.8).

L. Miklós et al., *Ecological Networks and Territorial Systems of Ecological Stability*, https://doi.org/10.1007/978-3-319-94018-2

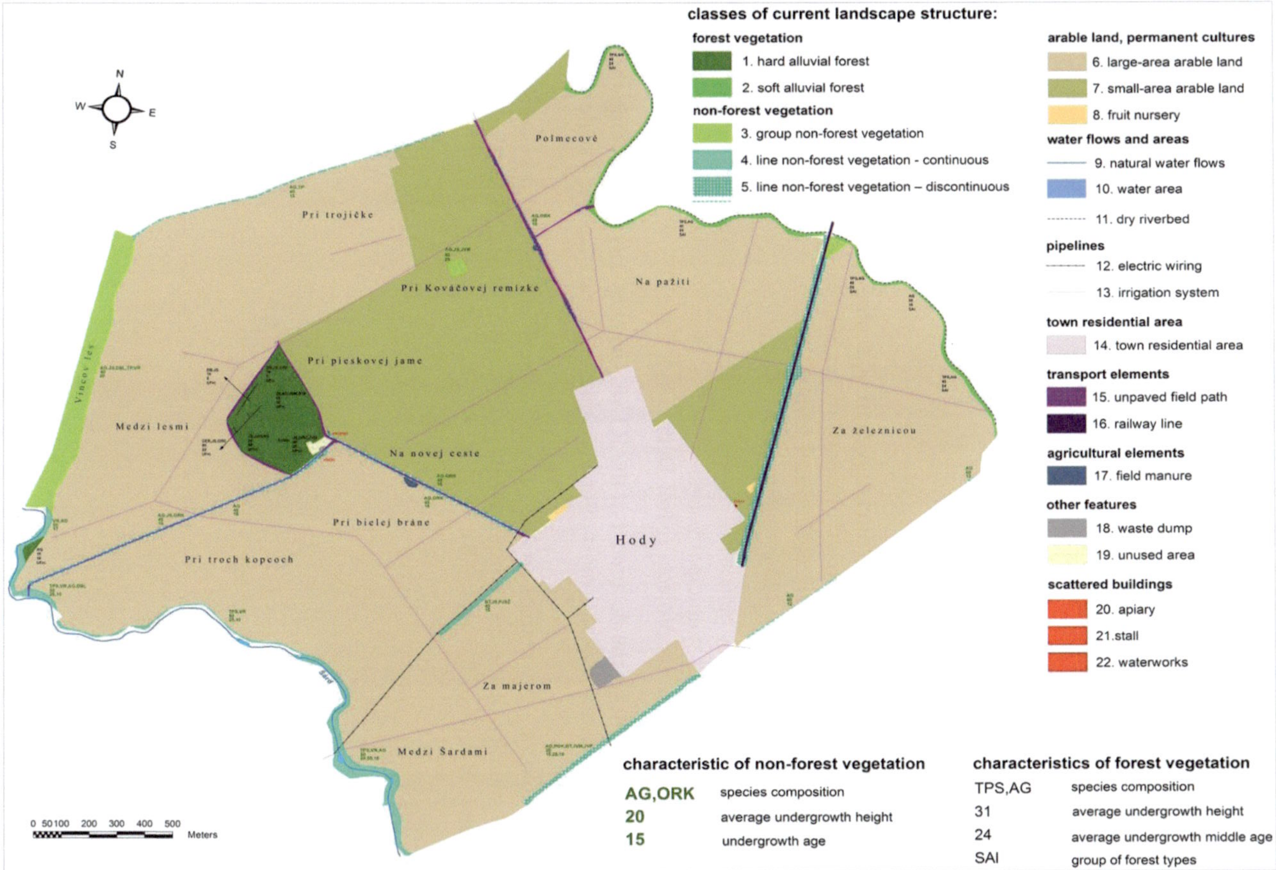

Fig. B.1 Secondary landscape structure/land cover

Fig. B.2 Abiocomplexes

Fig. B.3 Spatial synthesis of threatened factors

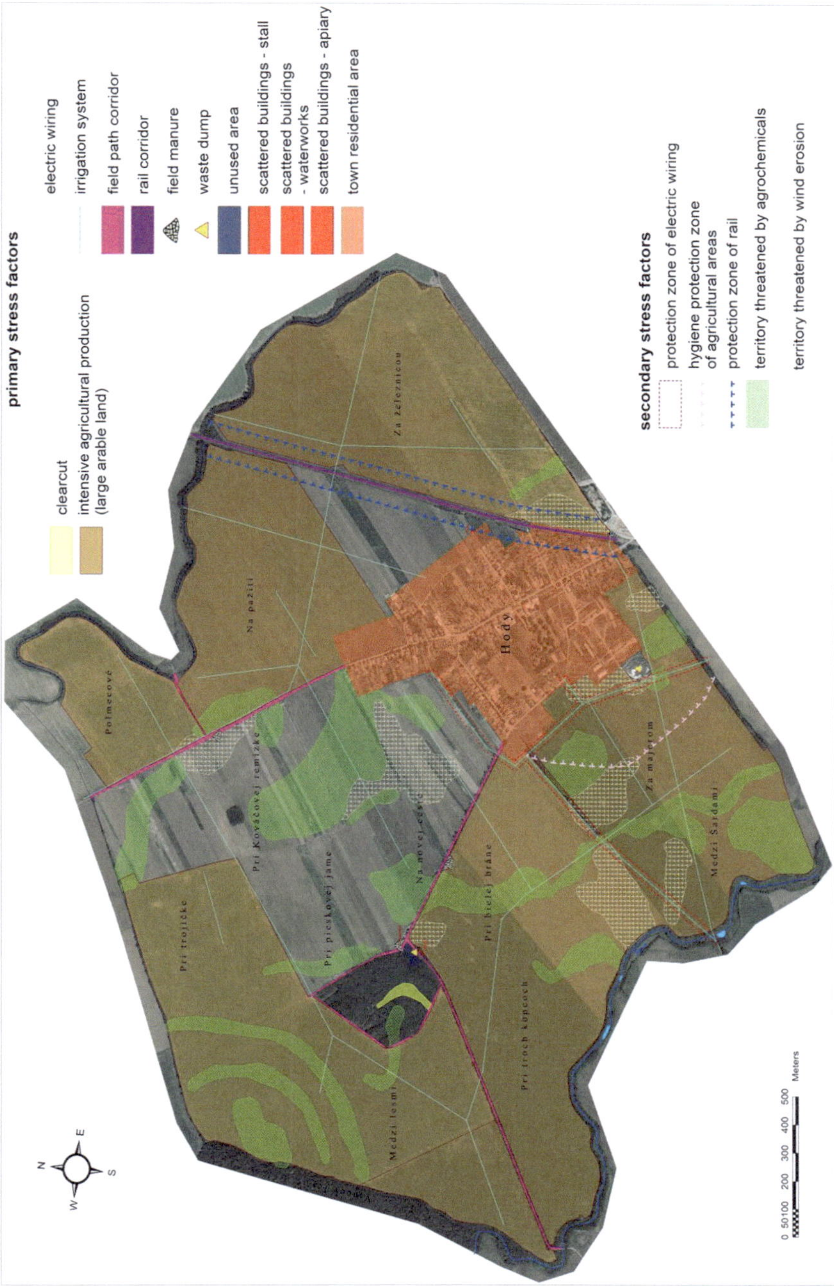

Fig. B.4 Spatial synthesis of threatening factors

Fig. B.5 Selected abiotic interpretation—trophism and cultivability

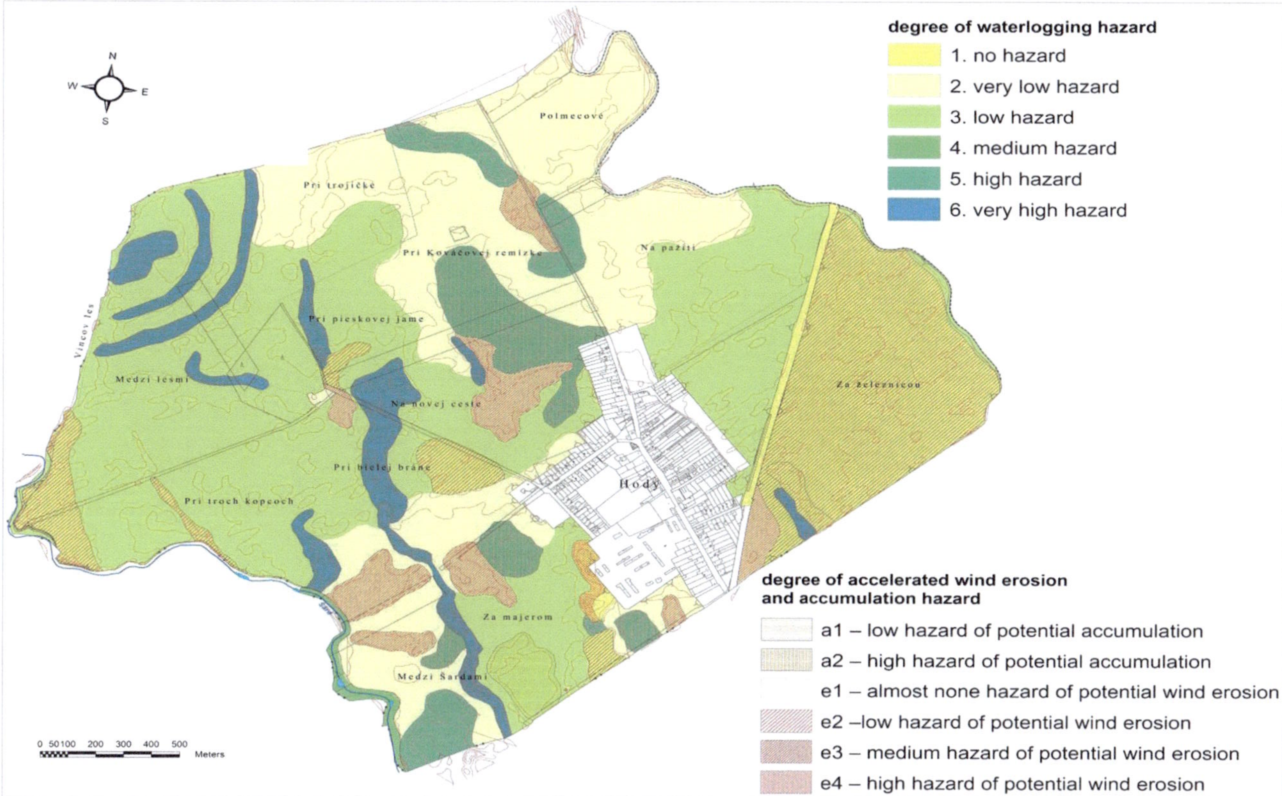

Fig. B.6 Selected abiotic interpretation—wind erosion and waterlogging

Fig. B.7 Classification of spatial ecological stability of the territory

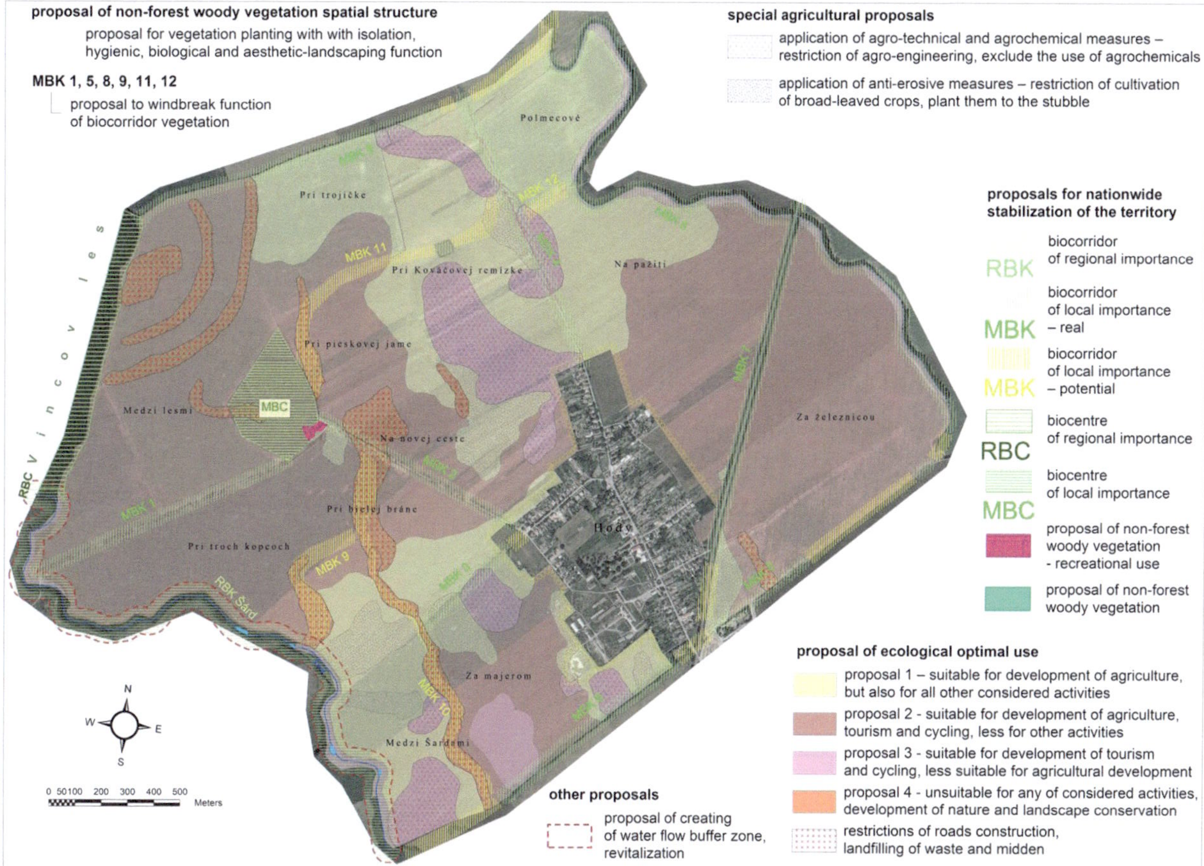

Fig. B.8 Proposals for the local territorial system of ecological stability

References

Diviaková A, Kočická E, Kočický D, Švec P (2005a) Miestny územný systém ekologickej stability k.ú. Galanta–Hody. Záverečná správa, ESPRIT, s.r.o, Banská Štiavnica.
Diviaková A, Kočická E, Kočický D, Švec P (2005b) Všeobecné zásady funkčného usporiadania k.ú. Galanta–Hody. Záverečná správa, ESPRIT, s.r.o, Banská Štiavnica.

Regulations and Conventions cited in the book:

Act No. 103/1990 Coll., amending and supplementing Act No. 50/1976. on Territorial Planning and Building Code (Building Act) as amended, and on amendments to certain laws
Act No. 127/1994 Coll. and No.24 / 2006 Coll. on environmental impact assessment
Act No. 134/2010 Coll., amending and supplementing Act No. 364/2004 Coll. on Waters and amending of the Slovak National Council Act No. 372/1990 Coll. on offences, as amended
Act No. 17/1992 Coll. on the environment, as amended
Act No. 211/2000 Coll. on free access to information, as amended
Act No. 237/2000 Coll., amending and supplementing Act No. 50/1976. on Territorial Planning and Building Code (Building Act) as amended, and on amendments to certain laws
Act No. 24/2006 Coll. on environmental impact assessment (EIA/SEA), as amended
Act No. 262/1992 Coll., amending and supplementing Act No. 50/1976. on Territorial Planning and Building Code (Building Act) as amended, and on amendments to certain laws
Act No. 284/1994 Coll. on nature and landscape protection
Act No. 3/2010 Coll. on the National Infrastructure of Spatial Information (NISI), as amended
Act No. 326/2005 Coll. on Forests (Forest Act), as amended
Act No. 330/1991 Coll. on ground arrangements, land ownership, land offices, land fund and land associations as amended
Act No. 364/2004 Coll. on Waters (Water Act), as amended
Act No. 384/2009 Coll., amending and supplementing Act No. 364/2004 Coll. on Waters and amending of the Slovak National Council Act No. 372/1990 Coll. on offences, as amended
Act No. 39/2013 Coll. on integrated pollution prevention and control (IPPC), as amended
Act No. 409/2014 Coll., amending and supplementing Act No. 364/2004 Coll. on Waters and amending of the Slovak National Council Act No. 372/1990 Coll. on offences, as amended
Act No. 50/1976. on Territorial Planning and Building Code (Building Act) as amended
Act No. 543/2002 Coll. on nature and landscape protection, as amended
Act No. 7/2010 Coll. on flood protection, as amended

L. Miklós et al., *Ecological Networks and Territorial Systems of Ecological Stability*, https://doi.org/10.1007/978-3-319-94018-2

Act No. 71/2015 Coll., amending and supplementing Act No. 7/2010 Coll. on flood protection, as amended
Council Directive 92/43/EEC on the conservation of natural habitats and of wild fauna and flora
Decree No. 24/2003 amending the Act on nature and landscape protection
Decree of the Ministry of Environment of the Slovak Republic No. 55/2001 Coll. on land-use planning materials and land planning documentation
Decree of the Ministry of Environment of the Slovak Republic No. 4/2003 Coll., implementing the Act. 543/2002 Coll. on nature and landscape protection, as amended
Decree of the Ministry of the Environment No. 492/2006 Coll.
Directive of the European Communities No. 79/409/EEC on the conservation of wild birds
Slovak National Council Act No. 3/2010 Coll. national infrastructure for spatial information
Slovak National Council Act No. 331/1991 Coll. on land consolidation and land ownership as amended (e.g. Act No. 549/2004 Coll.)

MIX
Papier aus verantwortungsvollen Quellen
Paper from responsible sources
FSC® C105338

If you have any concerns about our products,
you can contact us on
ProductSafety@springernature.com

In case Publisher is established outside the EU,
the EU authorized representative is:
Springer Nature Customer Service Center GmbH
Europaplatz 3, 69115 Heidelberg, Germany

Printed by Libri Plureos GmbH
in Hamburg, Germany